SLAY[...] SAVIO[...] SERVA[...] AND SEX

Springer
New York
Berlin
Heidelberg
Barcelona
Hong Kong
London
Milan
Paris
Singapore
Tokyo

SLAYERS, SAVIORS, SERVANTS, AND SEX

AN EXPOSÉ OF KINGDOM FUNGI

DAVID MOORE

DAVID MOORE
School of Biological Sciences
1.800 Stopford Building
The University of Manchester
Manchester M13 9PT
UK
david.moore@man.ac.uk
www.oldkingdom.org

Library of Congress Cataloging-in-Publication Data
Moore, D. (David), 1942–
Slayers, saviors, servants, and sex : an exposé of kingdom fungi / David Moore
p. cm.
Includes bibliographical references and index (p.).
ISBN 0-387-95101-6 (alk. paper)
ISBN 0-387-95098-2 (softcover : alk. paper)
1. Fungi I. Title.
QK603.M617 2000
579.5—dc21 00-059586

Printed on acid-free paper.

Production managed by Steven Pisano; manufacturing supervised by Jeffrey Taub.
Typeset by Impressions Book and Journal Services, Inc., Madison, WI.
Printed and bound by R.R. Donnelley and Sons, Harrisonburg, VA.
Printed in the United States of America.

9 8 7 6 5 4 3 2 1

ISBN 0-387-95098-2 SPIN 10773613 (softcover)
ISBN 0-387-95101-6 SPIN 10773833 (hardcover)

Springer-Verlag New York Berlin Heidelberg
A member of BertelsmannSpringer Science+Business Media GmbH

PREFACE

IF THERE WAS A MONSTER weighing over 150 metric tons rampaging through the forests of Oregon eating the trees, you should know about it, right? I mean, it's not like it's a new arrival. It's been there for the past 2,400 years. So why don't you know about it? What are they hiding from you? Maybe they don't want you to know that the beast has a body of hair-fine filaments and worm-like strands that extends over 890 hectares (2,200 acres), thrusting across the forest floor and reaching up to tear the life-wood out of the unsuspecting trees. No, maybe they don't want you to know about that. It could keep you awake at night.

I can understand. Trust me. I sympathize. You might lie there worrying that your legs could fall off. What? They haven't told you about that either? Not a word? About the terrible thing that ". . . consumed the people with a loathsome rot, so that their limbs were loosened and fell off before death . . ." (genuine eyewitness sound-bite). Unbelievable, isn't it? Silence in official circles at a time like this? Doesn't surprise me, though. Everybody knows about animals. Everybody knows about plants. But answer me this: who knows about fungi? You don't know because they are keeping quiet.

The fifth kingdom, that's what fungi are. More like a fifth column if you ask me; they're all around us, but nobody knows about them. How many fungi can you name? Two? Three? A dozen? I'm willing to bet that it won't be a large number. Yet there are more fungi on this planet than any other type of organism, and they've been here longer than anything else, too. We wouldn't be here if the fungi hadn't helped along the evolutionary way. Now we depend on them absolutely, day in, day out, for our very existence and our everyday pleasures.

So why are we so ignorant about them? Is it a government conspiracy? Well, maybe, maybe not, but I don't care, I'm going to blow the whistle; even if the black-clad agents of the Fungal Bureau of Investigation do come hammering down my door. It's time the truth was told. It's time the warning was given.

Don't turn your back on the mushrooms. . . .

CONTENTS

1

Toxins: Kill the Primates, Rule the World

Or, Don't Turn Your Back on a Fungus!

Like most little girls on their way to visit their grandparents for a Sunday treat, Innocenta probably bounced around in the back of the car like an overexcited rabbit. Just three years old, she was the only native-born American in the family. Her parents and grandparents had emigrated to southern Illinois several years before. One Sunday in 1995, after a recent move to the Gulf coast of Mississippi, they had a family get-together.

Grandfather particularly liked their new home because the wooded hillsides inland from the coast reminded him of home. He took readily to roaming through the woods and found many plants and mushrooms he recognized as being the same as those that he and his wife had collected for more than thirty years back in the old country, and he began collecting them again. (Amazingly, the locals didn't seem to be interested in this natural harvest.) For this special day, he found a good crop of succulent mushroom caps to collect, and his wife prepared them breaded and deep-fried as a novelty dish for her family's Sunday brunch.

The day went well until about 12 hours after the meal, when four of the five family members who had eaten the mushrooms experienced nausea, watery diarrhea, and abdominal pain and were rushed to the local hospital's emergency department. Grandfather did not have any symptoms and refused medical treatment.

The medical staff in the emergency department realized the possible significance of the meal of mushrooms and treated the family by giving them intravenous fluids and antiemetics so as to relieve their symptoms. Importantly, they also monitored liver function. When liver enzyme abnormalities were recognized the next day, all four patients were transferred to the medical center at the University of South Alabama.

Meanwhile, an expert on mushrooms from the State of Mississippi Toxicology Laboratory had been called to examine mushroom specimens collected by Grandfather. He examined them in detail and identified them as the Fool's Mushroom, which is known scientifically by the name *Amanita verna.* The family were all suffering from mushroom poisoning.

At the time they were transferred to the university medical center, all four patients had normal vital signs and were free of symptoms. But the possibility that toxins still coursing through their blood streams were doing irreparable damage to their livers caused continued medical concern. The function of the liver is to remove toxins from the blood, but by removing and concentrating them, it is itself in great danger of toxin damage. Therefore, liver function was closely monitored in all patients, and abnormalities were found that reached a peak about two to three days after that unfortunate meal. The patients were treated with intravenous fluids to keep their electrolytes in proper balance, with drugs to control bowel movements, and with activated charcoal to absorb and remove the remaining toxins from their guts to help their livers with toxin removal. The three adult patients were at their worst about two days after the meal but began to recover during the next two days.

Innocenta was not so lucky. Although she was free of symptoms three days after the meal, her liver was in rapid decline. Just one day later, she was transferred to a liver transplant unit where she was listed for an emergency transplant. On day five she required ventilator support, and on day seven she was removed from the transplant list when she developed pneumonia. She died from sudden liver failure and general infection on day 11.

This tragic story is likely to be repeated more frequently as time goes on. "Natural" foods have become much more fashionable, and their increasing popularity has already led to an increased incidence of mushroom poisoning among the general population; this incidence will probably continue to rise. Ironically, though, mushroom poisoning often occurs in experienced collectors. On the European mainland, amateur mushroom hunting is a popular pastime that has been practiced for centuries, yet this region has the highest reported incidence of mushroom poisoning in the world. In the United States, too, mushroom poisonings have been increasing since the 1970s in step with changing fashion. Not only are natural foods generally thought to be more wholesome, with wild mushrooms being especially promoted as gourmet delights in magazines, cook books, and restaurants, but the alternative use of mushrooms for their hallucinogenic effects has also tempted more people into mushroom collecting.

Yet the problem is not limited to those new to mushroom hunting. A Chinese magazine reported large-scale mushroom poisoning among villagers in Yunnan province in central China. Sixty-one people were working on a plantation on a hillside on June 28, 1997. During the day, 50 of them collected mushrooms for a communal meal in the evening. These were people who have lived with and used mushrooms throughout their lives, yet the mistake was made. Poisonous mushrooms were included in the meal. The first illness occurred at 6 A.M. the following day, and the first death, of a pregnant woman, occurred on June 30. By July 1, 51 of the villagers showed symptoms of food poisoning (10 people had been too busy to take part in the meal!), and a total of 26 of them died. Subsequently, *A. verna* and the death cap, *Amanita phalloides,* were identified as being in the collection used for the meal.

By all accounts, a very similar tragedy was narrowly averted in Italy in 1984 when a mycologist spotted the death cap (*A. phalloides*) on sale in Potenza market. He saved the town from catastrophe by making radio and TV broadcasts and touring the town by car, warning the locals of the danger by loudspeaker.

It is this close resemblance between poisonous and edible, even delicious, mushrooms that creates the problem. A little knowledge can be dangerous; even experts get it wrong. The 1991 edition of the authoritative *Larousse Encyclopedie* had to be recalled because the illustrations of deadly amanitas were labeled "indifferent" rather than "poisonous". By the time the error was noticed, 180,000 volumes had been distributed, and 250 people had to be hired to visit 6,000 book stores in France, Belgium, Switzerland, and Canada to overprint the correct wording on the right page.

In fact, fairly few mushrooms are poisonous to humans; fewer than 50 of the more than 2000 species of mushrooms that are commonly listed as occurring in most countries can be considered poisonous, and only about 6 are deadly. In a sense this point is irrelevant because unfortunately, most serious poisonings are due to the very species of *Amanita* that contain cyclic peptides called amatoxin and phallotoxin, two of the most potent toxins known. *A. phalloides* is aptly named death cap because it accounts for more than 90 percent of fatalities caused by mushroom poisoning in the United States and Western Europe, and the other two deadly species are *A. verna*, and *Amanita virosa* (the common name for which is the destroying angel).

Both of the toxins interfere with basic aspects of normal cell biology. Amatoxins bind to the enzyme RNA polymerase and interfere with production of the messenger RNA molecules, which take the genetic information from the genes themselves to the protein manufacturing machinery in the cell. With this process stopped by amatoxin, the cell is unable to produce vital working proteins. The cell cannot function properly, and so it breaks down and eventually dies. Body organs that normally have high rates of protein synthesis are particularly sensitive to this toxin. Liver cells (called hepatocytes) are in this group and are most commonly involved in mushroom poisoning, but other target organs include the kidney, pancreas, testes, and white blood cells.

Phallotoxin causes irreversible disruption of the cell membrane and cytoskeleton, and cell death inevitably results. Phallotoxin is not absorbed from the human gut, so is not thought to be responsible for the symptoms associated with human poisonings. Amatoxins are readily absorbed from the gut and are then concentrated in the liver where they do their damage. A patient poisoned with *Amanita* goes through four characteristic phases. For the first 6 to 24 hours the patient is free of symptoms; then there's a 12- to 24-hour period of severe abdominal cramps, nausea, vomiting, and profuse watery diarrhea. In this stage patients are often misdiagnosed as having gastric flu. In the third phase, as the patient's gastrointestinal symptoms improve over a further 12- to 24-hour period, this misdiagnosis may seem to be right, and the patient appears to be on the way to recovery. But all through this time the toxin is damaging the liver, killing its cells and literally ripping it apart from within. Symptoms of liver damage only appear four to eight days after the toxin was consumed, but the liver damage can progress rapidly, accelerating all the time. Death from catastrophic failure of the liver can occur anywhere from 6 to 16 days after the fatal meal was consumed.

If misdiagnosis can be avoided, if liver function is monitored, and if proper supportive care is made available, the patient has a fair chance of surviving the poisoning. In the long run, though, this may require a liver transplant. Even with the best medical care, death still occurs in 20 to 30 percent of cases, and the mortality rate of children less than 10 years old is more than 50 percent.

Those most at risk of being poisoned through eating wild mushrooms are toddlers, who are likely to eat anything on an experimental basis, recreational drug users, who think that any trip is worth a try, and immigrants, who are in danger of collecting toxic mushrooms that resemble species that are safe to eat in their home regions.

Evidently, cure is difficult and far from certain, so the best strategy is avoidance! Unfortunately there are no reliable rules to help you tell a poisonous mushroom from an edible one. One rule that you should obey, however, is take no notice of the old traditional advice that poisonous mushrooms make silver spoons turn black or that a mushroom that can be peeled is safe to eat. The most deadly poisonous of all mushrooms fail both of these so-called tests completely! There are a host of other, equally nonsensical, traditional ways of identifying mushrooms that rely on color and shape, smell and taste, milky exudates from the flesh of broken mushrooms or change in color of the flesh, ability to coagulate milk or change the color of onion or parsley during cooking. All are valueless in identifying poisonous mushrooms reliably.

The only certain test is eating, but don't put any greater reliance on those other bits of traditional "wisdom" that suggest that mushrooms showing evidence of being nibbled by rabbits or squirrels or eaten by slugs are edible. Slugs seem to be particularly fond of *A. phalloides,* and the rabbit digestive system is able to detoxify these mushroom poisons. In any case, while you are admiring the toothmarks around the margin of the dainty morsel you've just found, what makes you think the animal itself is not lying dead down a burrow somewhere?

I don't think we should put any faith in traditional wisdom relating to identification, and traditional methods of neutralizing the poisons in fungi by special methods of cooking are equally unreliable. From the time of the ancient Greeks, people have advocated boiling the mushrooms in oil, with meat, or with pear stalks. Vinegar is also supposed by many to neutralize the toxins. In the mid-nineteenth century, a French specialist experimented with a method involving steeping toxic mushrooms in vinegar or salt water for two hours, followed by boiling for half an hour. Repeated tests at the time gave some credence to the method, but it is doubtful that the original experiments really used the most toxic fungi, so the method is just as worthless as the rest. I've got to ask whether a mushroom, toxic or not, that's been pickled in salt water or vinegar and then boiled for half an hour is worth eating!

Nor do I have much confidence in traditional advice about identification or detoxification and, frankly, traditional treatments turn my stomach! One method that received much publicity at the time was based on the belief that a rabbit can eat the death cap without ill-effect. This leads to the conclusion that the rabbit's stomach must be able to neutralize the liver toxin and its brain must be able to neutralize the nerve toxin. From this logic came the treatment regime: take the stomachs of three rabbits and the brains of seven, chop them up finely, and give them (raw) to the patient mixed with sugar or jam. Bearing in mind that the patient will have been suffering from distressing vomiting and diarrhea, I suspect that a meal of raw rabbit stomach and brain would not have been a high point in the recovery process.

Mushroom poisoning is far from being a new problem. It is mentioned in ancient Greek and Roman writings. Pliny the Elder said that " Although mushrooms taste wonderful, they have fallen in disrepute because of a shocking murder. They were the means by which the emperor Tiberius Claudius was poisoned by his wife Agrippina" Toxic mushrooms are very convenient for poisoning, and attempts to use them have not been restricted to ancient Romans, though there are only a few cases on record. The most remarkable is a French case of 1918, in which a murderer who impersonated his victims to buy life insurance used various biological agents, including bacteria such as anthrax and typhoid, and toxic mushrooms to dispose of the victims. He was eventually caught because an insurance company doctor realized that the corpse bore no resemblance to the "insured" he had examined some time before the lethal event!

Although dramatic and tragic, death from mushroom poisoning occurs in a minority of cases. Most people who eat a poisonous mushroom suffer an unpleasant but relatively mild and short-lived bout of food poisoning. The quicker the symptoms arise, the less severe the attack. The toxins of deadly mushrooms are slow-acting and do not cause symptoms for 6 to 12 hours or more after eating. Those causing milder poisonings make people sick in two hours or less. After a few hours of discomfort (vomiting, nausea, and cramps, accompanied by diarrhea), the victim recovers completely.

The milder symptoms of mushroom poisoning are considered desirable by some people. For generations, in several areas of the world, intoxicating mushrooms have been eaten in connection with religious ceremonies. For example, some tribes on the Kamchatka peninsula in Siberia have eaten the fly agaric mushroom (*Amanita muscaria*) because the toxins it contains give rise to intoxication, hallucinations, and, allegedly, superhuman feats of strength. The Vikings are said to have used these mushrooms for the same purposes. But the most widely known mushroom-eating ceremonies are the magical sacred rites that originated with the Aztecs. The mushrooms used by the Mexican Indians are species of *Psilocybe, Conocybe,* and *Stropharia.* They contain psilocybins, which are chemicals that cause visual hallucinations, ecstatic states, altered perceptions of time and space, and, in a good "trip," general feelings of exhilaration and well-being that last for several hours. These are hallucinogenic fungi, and not surprisingly, mushrooms able to produce intense excitement in the consumer are now deliberately used for that purpose. Such psychoactive fungi, or so-called magic mushrooms, are eagerly sought by those yearning for the narcotic intoxication they offer. They are not uncommon fungi, being mostly specialized to the degradation of waste plant material. They prefer disturbed ground and so are often found around people—in parks, gardens, construction sites, as well as in forests and fields around the world. Fresh supplies can be found in most places, and can be bought in many. Indeed, walk around the center of Amsterdam and you will find well-appointed shops selling all of these, and *A. muscaria* as well. But there is no guarantee of a great trip! The psychotropic effect is a toxic effect. Different from the *Amanita* toxicoses, yes, but the semiconscious mild delirium and hallucinations are reactions of the nervous system to the toxins the psilocybin fungus contains. Whatever their protagonists may claim, there is no more enlightenment in fungal intoxication than there is in alcoholic intoxication; it's all a matter of the misfiring of a nervous system tortured by chemicals that it was not designed to cope with.

If you go out collecting wild mushrooms, the key is to be sure about what you collect for eating, but remember that *Amanita* fruit bodies are commonly mistaken for edible species, and cooking does not destroy the toxins. *Amanita* is widely distributed because it is mycorrhizal, a fungus that lives in symbiosis with the roots of a host tree, mainly birch, so it will be present in many woodlands. And these are the very places where you are most likely to go to find edible mushrooms. *Amanita* does not have a characteristic smell or taste, and colors vary with weather and soil conditions, as well as with the age of the mushroom. There are three features that do identify *Amanita:* the gills underneath the cap are white, there is a skirt-like ring, or annulus, around the top of the mushroom stem, and there will be an inverted skirt (a volva) arranged like a pouch around the swollen base of the mushroom. As it is much easier to detect the presence of the swollen base by digging the entire fruit body out of the ground, it is not good practice to simply collect mushroom caps or even cut off the stems at ground level.

Eating wild mushrooms can be deadly, and experience shows that even competent collectors can make mistakes, but overdramatic reaction is not necessary, nor is it helpful. I have seen suggestions that because of the risk of accidental ingestion, "the public should be advised to collect the fruiting bodies of mushrooms as they emerge from the ground and dispose of them." Indeed, a newspaper report of September 1998 told of the hysterical reaction at an English primary school when a child picked up a mushroom on the school playing field. The child was rushed to hospital, the school's annual sports day was postponed, and the field was sprayed with fungicide. At the end of the day the child left the hospital fit and well, and the local university identified the "deadly mushroom" as a common edible one! If this is the hysterical aversion end of the spectrum of reactions to mushrooms, I think the other end of the spectrum is equally ridiculous. This is the end populated (possibly temporarily in view of their habits) by dedicated mushroom collectors who tempt fate by attempting to distinguish between the fatal death cap mushroom, *A. phalloides,* and its edible look-alike *Amanita caesarea.* To me this is not sensible mushroom collecting; rather, it is an extreme sport such as free-fall parachuting, in which a major part of the attraction of the exercise is the very fact that you risk your life by doing it.

I do not have much sympathy with either hysterical aversion or deliberate danger-seeking. Eating one of the most toxic mushrooms can lead to death. Indeed, some people have suggested that eating just one cubic centimeter of one of these mushrooms is enough to kill the average person. Data on serious mushroom poisonings are not widely collected, so the

A toxic trio. From left to right, *Amanita muscaria, Amanita phalloides,* and *Amanita virosa. Amanita muscaria* is the cartoon toadstool, the fly agaric; it has a bright red cap with white dots on it so it's easily recognizable. *Amanita phalloides* is white with a greenish tinge, and *Amanita virosa,* appropriately, is deathly white. Photographs kindly provided by Mrs. Jo Weightman.

exact number who die each year from this cause is not known. It's been suggested that deaths due to mushroom poisoning in the United States are fewer than those caused by bee stings or lightning. This is probably not much consolation for the victims, but it is also not a logical reason for wanton destruction of the mushrooms in nature. In Britain, about 20 civilians are killed in road traffic accidents involving police cars each year. Should we make the police walk? My rule of thumb for eating mushrooms is only eat the ones with dark-colored (brown, purple, black) spores. You might miss a few delicacies that way, but you'll miss the toxic ones, too! Oh, and cross the road carefully when you hear police sirens.

Mushrooms are not the only fungi able to produce toxins that have plagued humanity over the years. Imagine a disease whose victims feel as if they are burning up, literally on fire, or as if ants, or even mice, are crawling about beneath their skin. Imagine that as the disease develops they may suffer terrible hallucinations, so bad that they sometimes drive the victim insane. If they avoid the hallucination and madness, an even greater suffering awaits. After a few weeks limbs become swollen and inflamed, violent burning pains alternate with feelings of deathly cold, and

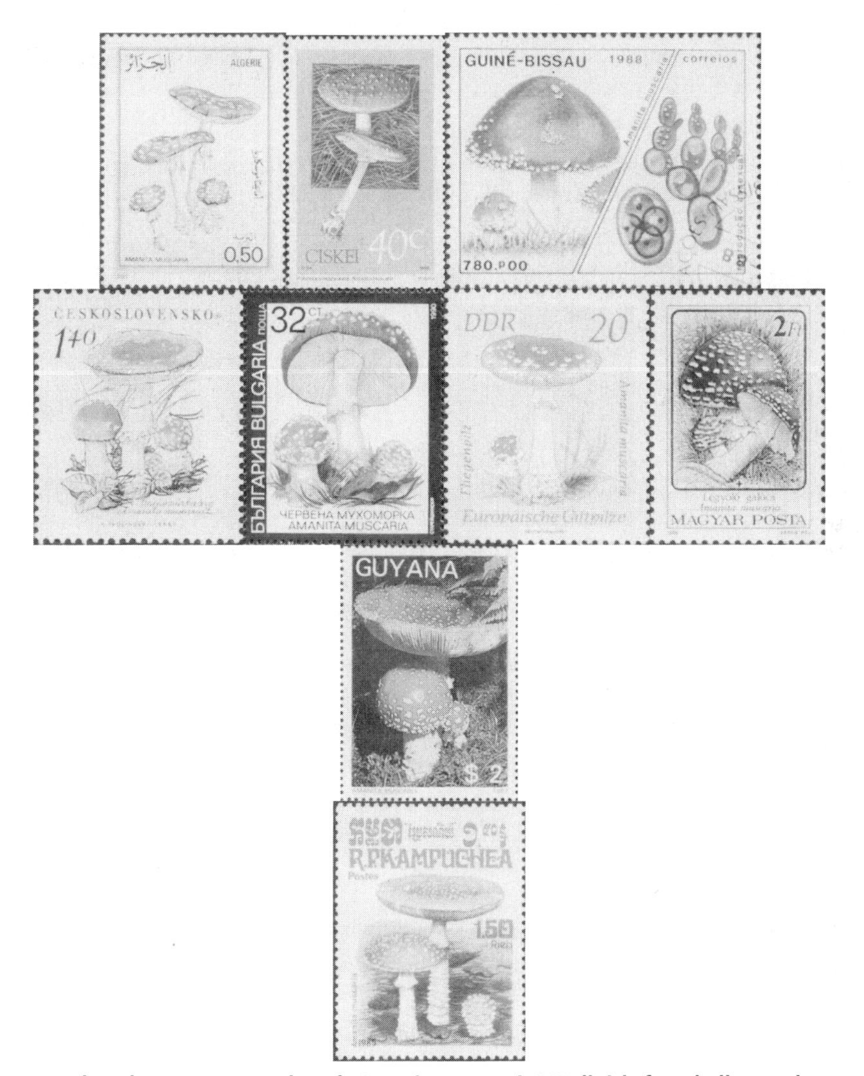

Remember that cartoon toadstool, *Amanita muscaria?* Well, it's found all over the world, and its appearance on postage stamps is testimony to this. Even this limited selection show that the fly agaric is truly international, being found in Africa (top row), Europe (next row), South America, and South East Asia (bottom row).

gradually the affected parts become numb and simply fall off the patient's still-living body. Fingers, toes, hands, feet, arms, and legs may all separate from the body and rot away.

The first record of this terrible disease dates from 857 A.D.; a chronicle from Kanten in the Lower Rhine region of Germany describes a ". . . great plague of swollen blisters that consumed the people with a loathsome rot, so that their limbs were loosened and fell off before death. . . ." At intervals throughout the following centuries there were epidemics in which an enormous number of victims died. One bad outbreak in France in 944 A.D. reputedly killed 40,000 people; the last epidemic of any considerable extent was in the 1880s. Records in the eleventh and twelfth centuries refer to this plague as holy fire, but it became associated with St. Anthony, as those suffering from the disease started to visit the saint's relics to seek relief and solace in faith. Most cases of what therefore became known as St. Anthony's fire occurred in France, and its distribution reflected the cultivation of rye and use of the grain to make bread. The human disease is caused by eating bread prepared from rye that is itself suffering a fungal disease. The plant disease, called ergot, is caused by the fungus *Claviceps purpurea,* which is parasitic on wheat, rye, and other grasses. Because it is most common on rye, it is often called ergot of rye. Spores of the fungus are carried by the wind until they land and infect the flower of the grass. The spore germinates on the stigma of the flower and then grows into the ovary. There it uses the food intended to nourish the seed to produce a curved mass of compacted cells, called a sclerotium, in place of the seed. When ripe, the sclerotium projects from the head of the rye as a hard purple to black slightly curved thorn. The sclerotia resemble the spur on the foot of some birds, which is called *ergot* in French; hence the common name of the disease. In wild plants, these sclerotia fall to the ground in autumn where they overwinter and germinate in spring to infect the next generation of plants. But on farms the sclerotia are harvested and milled with the grain. The flour then becomes contaminated with the toxins the sclerotia contain.

Ergot yields three groups of compounds, ergotamines, ergobasines, and ergotoxines, that have now become medically important. The first two groups excite smooth muscles (which control the uterus, blood vessels, stomach, and intestines) and are now used to promote contractions of the uterus in childbirth, to stem hemorrhaging, and to control migraines. The third group, ergotoxines, have an inhibitory effect on those parts of the nervous system that affect mood and emotional state and are used to treat psychological disorders such as delirium tremens and hysteria. The use of ergot as a clinical drug has been recorded from the sixteenth century, when the ergots themselves were used by European midwives to induce labor and to terminate unwanted pregnancies.

Ergot poisoning, now known as ergotism, causes two groups of symptoms that are manifestations of the same type of poisoning, gangrenous and convulsive ergotism. The former was frequent in certain parts of France, and the latter occurred mainly in Central Europe. The convulsive form results from ergot poisoning combined with a deficiency of vitamin A, probably itself caused by a lack of dairy produce in the diet.

In gangrenous ergotism, constriction of the blood vessels, especially to the extremities, eventually causes gangrene and death. In convulsive ergotism, the effects of the ergot alkaloids on the nervous system are most pronounced. Twitching, spasms of the limbs, and strong permanent contractions, particularly of hands and feet, occur. In severe cases the whole body is subject to sudden, violent, general convulsions, all coupled with hallucinations and visions. Ergotism has been suggested by some to explain the convulsive fits that took hold of eight young girls in Salem and led to the witch trials of 1692, during which 19 people, mostly women, were pronounced guilty and hanged.

Britain has been singularly free from ergotism. There is only one record of the typical gangrenous type, in the family of an agricultural laborer from near Bury St. Edmunds in 1762. The mother and five children all lost one or both feet or legs, and the father suffered from numbness of the hands and the loss of finger nails. Recognition of the disease, improvements in grain preparation, and especially the introduction of antibiotics in the latter half of this century to control the bacterial infections have all helped to remove human ergotism from our catalog of miseries. The only epidemic in the U.K. occurred in Manchester in 1927 among Jewish immigrants from Central Europe, who lived on bread made from a mixture of wheat and rye. All 200 suffered symptoms of mild convulsive ergotism. At the same time (September 1926 to August 1927) and perhaps because the grain came from the same source, there were 12,000 cases of convulsive ergotism in Russia. The last recorded outbreak, in France in 1951, caused more than 200 people to suffer from hallucinations, and 5 people died. One young victim believed she saw geraniums growing out of her arms. But ergotism in livestock remains a common problem arising from the ergot infection of fodder grasses. Sheep suffer from inflammation and ulceration of the tongue. In cattle, the symptoms are lameness and dry gangrene in the feet, and at times the whole foot can be lost.

There's one last fungal toxin that's worth mention, and certainly worth avoiding! It's called aflatoxin. As you might expect, there are several related aflatoxins, which are produced primarily by the molds *Aspergillus flavus* and *Aspergillus parasiticus*. They are considered to be the most active cancer-causing (carcinogenic) natural substances known. These toxins arise in crops like corn, peanuts, and to a lesser extent, rice and soybeans, even before harvest, but particularly when stored under the warm and moist conditions that permit growth of the infesting fungi. This is a problem in developing countries, especially throughout tropical Africa, and the southeast United States.

When the tainted crop is eaten, the aflatoxin is absorbed and metabolically activated by liver enzymes. Activated aflatoxins interact with the workings of liver cells leading to their death, which can cause hepatitis, or transformation into cancerous growths. The human liver has a fairly slow metabolism, so susceptibility to acute poisoning is relatively low. Animals are more in danger, particularly poultry. In the 1960s 100,000 turkeys died on poultry farms in England. The disease was unknown at the time and first became known as turkey X-disease. It was eventually found to be due to contaminated poultry feed, and this costly event initiated aflatoxin research. More serious to humans is that prolonged exposure causes primary liver cancer, as well as other cancers.

Aflatoxin health risks differ considerably around the world. Warm, humid climates are generally optimal for mold growth, and with poor regulatory and control systems, countries like Kenya, Swaziland, Uganda, China, and Thailand may have as much as 500 times higher incidence of liver cancer than is common in developed countries. The European risk comes mainly from importing of contaminated foods (mainly nuts) and animal feeds. Aflatoxin can enter the human food chain from such contaminated feeds. Grains, peanuts, other nuts, and cottonseed meal are among the foods on which aflatoxin-producing fungi usually grow. Meat, eggs, milk, and other edible products from animals that consume aflatoxin-contaminated feed are also sources of potential exposure to humans. To protect against too much exposure, most developed countries measure aflatoxin levels, and acceptable levels in human food and animal feeds are controlled by legislative authorities. A number of agencies regulate aflatoxin levels in the United States, but still, Americans may consume up to half a microgram of aflatoxins every day, and aflatoxins are among the more than 200 environmental chemicals that have been found in samples of human breast milk. The levels of hazardous chemicals like this in human breast milk are a measure of our progress (or lack of it) in cleaning up our environment. One expert says that "there is no uncontaminated mothers' milk anywhere in the world . . . all mothers carry environmentally-derived chemicals in their bodies." Nevertheless, the benefits of mothers' milk to a baby are incalculable because the baby is provided with antibodies against infection, protection against allergies, and particular nutrients vital to intellectual development. So although the levels of toxins found in breast milk are a cause for concern, there isn't any hard evidence yet that they can harm the baby, and the advice is still that most women should breast-feed. The emphasis is shifting toward avoidance of the toxic chemicals. On the one hand, mothers-to-be could avoid meats and freshwater fish that could be contaminated; this usually means avoiding the cheaper products because corners could have been cut to make them cheap. Reduced consumption of dietary fat is also a good idea because many toxins are accumulated in an animal's fat.

On the other hand, the industrial watchdogs and regulators can control the basic crops that are at risk to minimize contamination and prevent entry of contaminated materials into the food chain. Crop rejection and losses associated with laws regulating the aflatoxin contents have significant economic impact, primarily but not only on developing countries. Costs caused by aflatoxin contamination of corn, for example, were estimated to be about 240 million U.S. dollars annually throughout the 1980s. The United States poultry industry losses from aflatoxin poisoning greatly exceed 100 million U.S. dollars per year. Most of the loss results from slow weight gain and reduced feed efficiency rather than from mortality.

Control of aflatoxin is restricted to physical separation of contaminated lots, partial decontamination, and prevention of further contamination by provision of good storage conditions. Procedures that try to remove aflatoxins are generally inadequate, so the best chance of making sure that foods and feeds are free of aflatoxins is by preventing mold infestation and toxin production in the first place. Condensation of moisture on roofs and walls, leaking roofs, and seepage of water into warehouses and the like are some of the causes of mold growth during storage. Every place where moisture can get to the product and cause mold growth must be eliminated.

Refrigeration and climate control are useful. Reducing the oxygen concentration (to less than 1 percent) and increasing the carbon dioxide (to above 20 percent) can also inhibit mold growth. Sadly, all of these measures are costly, and since huge tonnages of aflatoxin-contaminated produce come from developing countries in the tropics, most of these desirable storage measures are completely impracticable. We have to fall back on continued vigilance. Rely on this: the mold will be there. Don't drop your guard!

These fungal toxins are so effective at killing, maiming, and disorienting humans that it is worth asking why the fungi produce such deadly toxins. Among the mushrooms and toadstools, they are fairly obviously protective agents, defending the mushrooms against destruction by grazing. The mushrooms are structures intended for spore distribution. Their shape protects the spore distribution mechanism from the rain. The toxins protect the mushroom against animals that might otherwise eat it. An eaten mushroom is not a successful mushroom! My argument about the purpose of toxins is an evolutionary one. I interpret toxins as products that are intended to improve the evolutionary fitness of the mushroom. Survival of the fittest is the basic Darwinian statement. What it means is that the individual that is best suited will survive to produce progeny. The crucial point is that the Darwinian selection process must benefit the individual. The benefit to the mushroom is that it should not be eaten but should survive to distribute its spores. Yet the amanitas are fairly anonymous species; people die because they mistake the *Amanita* for some innocuous mushroom. The toxin is not a warning because there is no accompanying warning coloration that the eater could learn to avoid on some future occasion. So I believe the toxin is an immediate deterrent. Something which will cause immediate distaste, slight injury, or a jolt to the nervous system which will make the eater stop eating and go elsewhere. This is not a matter of flavor; it's more the impact of the chemicals on the eater. Despite the lurid title of this chapter, I don't think the toxins are aimed at large mammals like humans. A large animal will eat the whole mushroom; then both will end up dead, the animal will be killed by the toxins, and the mushroom—with its load of progeny spores—will be digested in the stomach of what will by then be a rapidly rotting corpse. No, I think the toxins are aimed at small animals. They are chemicals that make the fungus tissue immediately distasteful to insects, slugs, and small mammals that might chew away small chunks of the mushroom at a time. Animals of this sort are not intellectual heavyweights. There is no place for subtlety here, so a very potent toxin (the higher the potency the less the mushroom will need to produce) that effectively kicks the grazer in the mouth will do the job very nicely. Poisonous fungi have enzymes similar to those the toxins attack, but the fungal enzymes are more resistant to the toxins than the same enzymes of animals.

This could certainly also apply to ergot. Remember that the ergot sclerotium must outlast the winter to perform its biological function of reinfecting plants in the spring. The toxins it contains may well deter soil animals from eating too much of the ergot during its winter of waiting. The same argument is applicable to aflatoxins because the molds that make them are in competition with mites, weevils, and insects for the food value that the stored crop represents. Producing a toxin powerful enough to drive away these competitors could be an important way of tipping the balance in favor of the fungus. Unfortunately for us, such potent toxins also kill the primates.

I've seen it suggested that the simple answer that toxins protect against being eaten is not very satisfying. The argument runs that if this feature were so important, poisonous mushrooms would be expected to be more abundant than the innocuous ones and eventually would be expected to become the dominant type (which has clearly not happened). This line of argument is too silly to take seriously. The point is that no one is suggesting that producing the toxins is the only, still less the best, survival strategy. It is one of a number of potential strategies. Some organisms may take advantage of the possibility that being eaten helps disperse the spores (spores are not easily digested and may pass through the intestines). Others may strengthen their cell walls to become too tough to eat, or adapt to become inconspicuous. They are all valid evolutionary strategies, and they all contribute to life's rich pageantry.

2

Blights, Rusts, Bunts, and Mycoses

Tales of Fungal Diseases

The notion that food is the driving force of human history is gaining acceptance. Finding enough to eat today so that you stay alive tomorrow is the mark of individual success. Finding so much today that you can afford to devote tomorrow to thinking is the start of civilization. The argument emerges that tools were devised to improve some aspect of food gathering, so the origins of technology are also to be found in the basic need to eat.

The main concern of early humans was clearly food gathering, and in that exercise, fungi most probably featured just like any other bounty of nature. Mushrooms are likely to have been collected for food alongside any other fruits and berries. I'm also very sure that the ability of yeasts on fruit left to one side to ferment the stuff into an alcoholic soup was a very early discovery indeed! But an early philosopher of even genius proportions would never have been able to recognize the influence of fungi in this miraculous event. However, when those early humans gave up their nomadic hunter–gatherer existence and turned to agriculture to solve the food problem, they would rapidly have been challenged by the fungi. Early farmers must have learned very quickly that crops are very uncertain resources, prone to variations in weather, fire, floods, weeds, insect pests, and those various sorts of plant disease that came to be referred to collectively as *blights*.

Great crop losses can be suffered, caused by any of these factors, but by bringing the crops together into fields in the first place, the early agriculturalist created ideal conditions for the spread of plant disease. And the more selective the farming was, the closer the crops came to being true monocultures, and the greater the extent of agricultural losses due to any single agency like a particular plant disease might be.

Standing out among the examples of how damaging a crop disease can be is the Irish famine of 1845–46, which was caused by the failure of the potato crop in Europe because of just one plant disease, the potato late blight. This is an astonishing story of how a crop disease affected the structure of our civilization and our understanding of nature, while causing the deaths of one in eight of the Irish population. It is a story that goes far beyond statistics of number of deaths due to starvation, number of people emigrating, or the number of crop losses and a reduction in agricultural yield, and I will tell you that story in some detail. But it is a piece of our history that we must read about with the knowledge that even today, world agriculture suffers significant losses due to plant disease, despite all our scientific advances since the mid-1800s. Hopefully, in that time we have learned enough at least to avoid massive calamities like the Irish famine. Today's losses can be reported in terms of monetary losses, yet behind each such statistic there must be personal tragedies in which the lives of individuals and families are changed dramatically.

The greatest losses suffered by agricultural crops today are caused by insect damage and plant diseases, with plant disease being the most damaging. Of course, factors other than fungi cause disease, such as bacteria, viruses, and even nematode worms (eel worms). There are serious plant diseases caused by all these other pathogens, but fungi probably cause the most severe losses around the world. For one thing, there are more plant pathogenic fungi than there are plant pathogenic bacteria or viruses. One survey taken several years ago in Ohio came up with the estimates that in that state, 1,000 diseases of plants were caused by fungi, 100 were caused by viruses, and only 50 were due to bacteria.

Agricultural survey statistics make it clear that crop losses directly attributable to fungi are very considerable. Of course, the statistics are constantly changing because they at least partly depend on the weather, but it appears that world agriculture sustains average losses (in terms of monetary value) of around 13 percent annually as a result of plant diseases. This overall average conceals instances of good news: disease loss in the 1 to 2 percent range; as well as bad news: a season of unusually heavy disease incidence that might involve losses in the 30 to 40 percent range. This is the agriculture of the present we're talking about, not some primitive agriculture of the distant past. Today, on average one in every eight crop plants will fail to yield because of fungal disease. That's a terrible tax on our activities to take into the shiny new millennium!

We lose more than mere money, too. A disease of the native American chestnut, chestnut blight (caused by an introduced parasite), effectively eliminated a stately and valuable timber and nut-crop tree from the United States. A similar loss, equally difficult to quantify because it was a loss of amenity as much as monetary value, happened in England when large elm trees were killed by Dutch elm disease (also caused by an introduced parasite, though this time the introduction was from the United States and into Europe).

All groups of fungi from the most primitive to the most advanced can cause serious plant diseases. For example, late blight of potatoes and downy mildew of grapes are diseases caused by the most primitive of fungi (so primitive that many experts would not include them as true fungi at all), whereas rusts and smuts are diseases caused by members of the group of fungi that is the most advanced in evolutionary terms. Chestnut blight, peach leaf curl, Dutch elm disease, net blotch of barley, beet leaf spot, apple blotch, maple leaf spot, and thousands of other diseases are caused by all those fungi in between these extremes.

Talking about extremes brings us back to the potato murrain. The word *murrain* is not in common usage these days; in fact, my dictionary describes it as being obsolete. It's the word that was used at the time, and the word stayed with the story as it passed into history. A murrain is a great pestilence, a disease so widespread that it becomes a plague. The story I'm going to tell you deals with just such a terrible disease of the potato crop, and it should come as no great surprise, considering what the rest of this book deals with, that it's a disease caused by a fungus. Before developing the story, though, it's essential to put it into historical scientific context. The events mostly took place in 1845 and 1846. This was approximately 25 years before the great French microbiologist Louis Pasteur proposed his theory that diseases (of humans and animals) were caused by "microbes" or "germs." Scientists in the mid-nineteenth century had little concept of disease. Individual beliefs and prejudices may have then determined one's convictions about the cause of the affliction: possibly a judgement from God, the work of the devil, that witch down the lane, or even the man next door with the evil eye. Nobody at that time would immediately associate dying plants with an infection. If such thoughts are not part of your everyday worldview, whether you are the king, a scientist, or a pauper, then just what can you do when all the plants in your fields die?

The first report of what was to develop into such a calamity can be found in a letter from Dr. Bell Salter in the Isle of Wight to *The Gardeners' Chronicle and Agricultural Gazette*. In the edition dated August 16th, 1845, Dr. Salter reported the arrival in England of a new malady of the potato. Matters progressed so rapidly that the distinguished editor of that journal, Dr. John Lindley, published in his editorial the next week, on August 23rd, that "A fatal malady has broken out amongst the potato crop. On all sides we hear of the destruction. In Belgium the fields are said to have been completely desolated. There is hardly a sound sample in Covent Garden Market."

The potato crop, just like other crops, had been affected before. It was damaged by too much rain in wet seasons or by too little rain during a drought. Sometimes the tubers were scabby and inferior quality, and sometimes the leaves curled up and the crop was reduced. But nothing as destructive as this new murrain had ever been seen before. Growing plants (remember this was in August, the summer season) looked like they had been badly affected by frost. Lindley's editorial told the story: "The first obvious sign is the appearance on the edge of the leaf of a black spot which gradually spreads; the gangrene then attacks the haulms (stems), and in a few days the latter are decayed, emitting a peculiar and rather offensive odor. When the attack is severe the tubers also decay." Lindley recognized that if this "gangrene" continued to spread, an important part of the country's food supplies for the coming winter would be lost. He offered little hope for treatment, though, saying, "As to cure for this distemper there is none. One of our correspondents is already angry with us for not telling the public how to stop it; he ought to consider that Man has no power to arrest the dispensations of Providence. We are visited by a great calamity which we must bear."

But who would bear the greatest calamity? As August became September, reports of the spread of the murrain came from Poland, Germany, Belgium, France, and from all over England. Lindley's fears were not exaggerated. Every strain of potato in cultivation was attacked. When they were dug from the field, the potatoes were blotched with the dark patches of rotting flesh that were symptomatic of the disease. They were covered in patches colored like bruised and battered human flesh, patches that smelled of pestilence. This was not something that ended when the crop was lifted. Potatoes left on the floor of a barn for a few days became worse than when they were lifted. This was the unique character of the potato murrain. It spread among potatoes in the ground and also among those in store. The crop you might have thought was safely harvested and stored could rot away in a few days; every tuber, no matter how slightly affected, would be lost. There is a record of a shipment of potatoes sent, routinely, by ship from the East coast seaport of Kingston-upon-Hull to be marketed in Belgium. Two thousand tons of potatoes left Hull in sound, palatable condition, but the entire cargo was rotten before the ship reached its destination just a few days later.

Then, on September 13th, Lindley made the most dramatic announcement in *The Gardeners' Chronicle:* "We stop the Press, with very great regret, to announce that the Potato Murrain has unequivocally declared itself in Ireland. The crops about Dublin are suddenly perishing . . . where will Ireland be, in the event of a universal potato rot?"

The potato murrain affected the whole of Europe, but right from the very start, it was clear that the Irish peasants would suffer most. Loss of the potato crop in England would bring distress, of course, but the poorest laborers in England lived on oat gruel and bread in addition to potatoes. The cereal crops were intact, so a major proportion of their normal diet was still available (no doubt at higher price due to competitive pricing strategies). In Ireland, on the other hand, the poorest members of the population lived almost exclusively on potatoes. On the average, 4 to 7 kilograms of potatoes were consumed per day, per person, day in, day out, for 10 months of the year. Over 100 kilograms were consumed per family per week. The only break in the cycle occurred in July and August, the gap between the old and the new potato crops, when the population had to subsist on wheat meal and anything else they could find. The population of Ireland had grown from 4 million in 1800 to over 8 million in 1845, largely because of the ability of the potato plant to provide large crops of easily stored tubers from small holdings on even relatively poor land. One acre of average land could produce 6 tons of potatoes each year. The tubers were easy to store over winter, so the potato had brought reliability of food supply to the poorest peasant farmers. It was clear to everybody that if the potato murrain spread through the small-holdings of Ireland, there would be millions of men, women, and children who would starve to death. And spread it did.

The crops of 1845 and 1846 were lost. In 1847 there was good weather, and although the amount of potato acreage was very much reduced in Ireland, the murrain was localized and relatively unimportant. With brilliant sunshine from July to September that year over the whole of Europe, harvests of grain and potatoes were good. The two worst years of famine were past, but the misery went on in the shattered Irish countryside. Two years of famine were devastating. In the years from 1845 to

1860, 1 million people died in Ireland as a direct consequence of the famine, and over 2 million emigrated. Those that survived changed the world. Many headed for North America, and the "Paddies" became the labor force that built the foundation of the United States. They brought the Catholic Church to a position of prominence in a nation founded by Protestants. Twenty percent of the U.S. population presently claims Irish ancestry. There's an estimate that in the 100 years between 1830 and 1930, 50 percent of those born in Ireland left the country to make their permanent home elsewhere. In 1996, Ireland, becoming a "tiger economy" within the European Union, reported its first net gain in population since the famine. The social upheaval caused by the Irish potato famine changed the demographic and political structure of the whole world and made an enormous contribution to the structure of the socioeconomic civilization we enjoy today.

What exactly was the potato murrain? The main intellectuals of the day were unable to come up with a cause, much less a cure. There was an official Commission of Enquiry that recommended some perfectly sensible methods of storage for sound potatoes. However, the crop then being harvested was not sound, and the tubers continued to rot in the stores. It's easy to be dismissive in retrospect, but at the time it was difficult to see what other advice the government's scientists could offer. Probably, for the time, the best the government could have done is what it did. Prime Minister Peel requested a survey of the state of the Irish potato crop. In response to this, on September 16, 1845, the Inspector General of Constabulary, D. McGregor, issued a strictly confidential circular from the Constabulary Office in Dublin Castle that started: "Information having reached the Government that the POTATO CROP of the present year has totally failed, from disease, in many districts of this Country, County and Sub-Inspectors of Constabulary are hereby directed to make full and immediate inquiries respecting the state of this Crop in their several Districts, and to report the result of such inquiries without loss of time . . . These inquiries are not only to be regarded as confidential but they are to be so conducted as to prevent speculation on the possible motives for seeking the information required." Over 200 of these disease reports had been received by the end of September. The extent of the disease was quite clear. The impact was also evident. Sir William Wilde wrote in the *Dublin University Magazine* of 1854 that "The late Bishop Brinkley, one of the most profound thinkers we have ever had in Ireland, who predicted the loss of the potato many years ago, and calculated mathematically the extent of ruin which was likely to follow, declared to his relative, the late Dr. Graves, that he was unable to sleep for an entire night, owing to the effect that it had upon him." From conscientious gathering of information, the government knew that about 2 million people were dependent entirely on the potato for food in Ireland, and 8 million pounds sterling were put into a relief effort in the 18 months between 1845 and 1847. That is about one pound sterling for every person in Ireland at a time when the agricultural laborer earned about two pounds sterling per year. A colossal amount for the day, though the tragedy was so great that even this relief effort was overwhelmed.

There were some hints as to the real cause of the famine. The peculiar changes of weather that had occurred during the summer of 1845 must have had much to do with the outbreak and spread of the potato murrain. The early part of the season that year was good: good for planting and good for early growth, so the crops looked very promising right up to July. In July the weather was hot and dry. The temperatures were up to 4 degrees above the average for previous years. F. J. Graham recorded all this in an essay on the history of the murrain, which won a prize and was eventually published in the *Journal of the Royal Agricultural Society*. He went on: "It then suddenly changed to the most extraordinary contrast that I ever witnessed in this fickle climate, the atmosphere being for upwards of three weeks one of continued gloom, the sun scarcely ever visible during the time, with a succession of most chilling rains and some fog, and for six weeks the temperature was from 1.5 degrees to 7 degrees *below* the average for the past nineteen years."

Dr. Lindley theorized about the cause of the murrain, but his explanation was wrong. Lindley argued that the cold and wet July weather caused the potato plants to become overladen with water. The man who got the explanation right, at the time, was the Rev. M. J. Berkeley who was, in Lindley's words, "a gentleman eminent above all other naturalists of the

United Kingdom in his knowledge of the habits of fungi." Berkeley associated the murrain with growth of a kind of mold in the affected tissues. He saw the damaged foliage himself in his parish, near King's Cliffe in Northamptonshire, and put forward the revolutionary theory that the mold was the cause and not the consequence of the potato murrain. That the murrain, in other words, was a disease.

Lindley argued that Berkeley was taking his interest in fungi too far, and that any mold that might be present was one of the "myriads of creatures whose life could only be maintained by the decomposing bodies of their neighbors." The way in which molds and mildews appeared almost overnight and reproduced was a great mystery at the time. Felice Fontana had examined rust on wheat under a microscope in 1767, recognizing it as a minute vegetable with bodies resembling seeds. Indeed, in 1807 in France, Bénédict Prévost had actually seen the spores of the Bunt fungus on wheat germinating like seeds in water. Nevertheless, it was still commonly believed that small fungi could be produced in decaying matter by spontaneous generation. Robert Hooke published his *Micrographia* in 1667. He was the first person ever to use a compound microscope to describe the appearance of fungus growth. He examined a blue mold on some old leather and some rose leaves that had yellow spots on them. His opinion was that "the blue and white and several kinds of hairy spots, which are observable on different kinds of putrify'd bodies are all of them nothing else but several kinds of small and variously-figured mushrooms, which from convenient materials in those putrefying bodies, are by the concurrent heat of the air, excited to a certain kind of vegetation".

The belief of the day, therefore, was that putrefaction caused the appearance of the lesser organisms. The Rev. M. J. Berkeley was going against this core belief by making his suggestion. But he was in frequent correspondence with other naturalists in Europe, and from their observations, he was able to confirm that the same mold fungus was associated with this potato disease across the Channel. The same mold was on the diseased potatoes themselves. Though Berkeley had never seen this particular species of mold before, it resembled one he had seen growing on onions and shallots. It also appeared to be related to the fungus that was associated with a very serious disease of silkworms in Europe. The potato murrain, the Rev. M.J. Berkeley insisted, was due to the growth of this particular fungus, and no other, as a parasite on the potato plants. Berkeley called it a species of *Botrytis infestans;* it is now called *Phytophthora infestans,* the causal organism of the late blight of potato. The cool, wet, and very humid weather at the end of July favored the spread of this moisture-loving fungus, and the disease became an epidemic.

There is great scientific significance in Berkeley's ideas as to the most probable cause of the potato murrain. His was a new conception of the nature of disease. By advancing the hypothesis that a living parasitic organism on the potato foliage was the cause and not the consequence of the potato murrain, the Rev. Berkeley was doing no less than anticipating the germ theory of Pasteur by almost 25 years. Of course, Pasteur proved it, and Berkeley only suggested it. However, it's a measure of the intellect that was at work in King's Cliffe in Northamptonshire late in the summer of 1845.

Sadly, despite all the varied intellects that were devoted to the potato murrain, no cure or treatment emerged. We will never know what twists human history might have taken if some person had remained alive at some stage, or some other person had succumbed. We will never know what human genius was withered away before full bloom, caught in the tragedy of illness, famine, or mortal accident. But we do know that in the last weeks of July 1845, a series of events occurred that were to change the whole course of human history. And the agent was a fungus.

A sad postscript to this story is that one person of the day did have an effective way of saving the potato crop. It was widely known that the murrain started on the foliage of the potato plant and then made its way to the tubers. Dr. Morren in Belgium, within two weeks of the first appearance of the murrain in 1845, pointed out that if the upper parts of affected plants were removed and destroyed, the tubers remained healthy. Small, perhaps, but healthy. If Morren's measure had been adopted, much of the potato crop could have been saved in the famine years.

There is just one other plant disease that I will discuss in some detail because of the general points that it covers: Dutch elm disease. This tree disease was first noticed in Holland in 1918 and 1919. Elms died soon after first showing symptoms, and mature trees were lost in large numbers. For most people the immediate aftermath of the First World War in Europe was a more immediate concern rather than this disease of elm trees. But there were numerous ideas as to its cause, including that it was a side effect of poison gas from the war front. Dead and dying elms were not confined to sites where these fanciful explanations might apply, however. It came to be called Dutch elm disease because its true cause was established at that time by Dutch scientists. They found it was caused by a fungus with two different types of spores. Today we know the fungus as *Ophiostoma ulmi*. It was actually brought to Europe from the Dutch East Indies in Southeast Asia during the late nineteenth century. Wilting, caused by lack or water and nutrients, is a symptom of Dutch elm disease. First, leaves droop and turn yellow. In a few days to weeks, the leaves are brown and dead. Larger branches begin to die, and most trees are completely

Late blight is still with us and is still able to destroy potato crops in the field and rot them in store. The storage bin on the left was filled to the brim with freshly harvested potatoes. But they had been infested with late blight, which turned them into the festering mess shown in the photograph on the right. Photographs kindly provided by Dr. H. A. S. Epton, School of Biological Sciences, The University of Manchester.

dead within two years. The reason for this progression is that the fungal cells plug the channels that normally distribute water and nutrients around the tree. As more channels are plugged, the diseased tree is starved of food and water. When the tree dies, the fungus will grow on the wood that remains. However, there is more than a fungus involved. Left to itself, the fungus has difficulty passing from tree to tree. In this case, the factors that turn the disease into a raging epidemic is a relationship between the fungus and several species of elm bark beetles. Adult females of these insects lay eggs in recently dead elms. Their eggs hatch, and young larvae tunnel into the inner bark and outer wood to feed on it. If the tree has been killed by Dutch elm disease, the fungus sporulates in the beetle tunnels so the adult beetles that emerge are covered with fungus spores. These are transported to the first tender, young, healthy elm twigs that the young beetles bite into. Elm bark beetles, therefore, are vectors of Dutch elm disease.

Chinese and Siberian elms are highly resistant to the disease, but elms native to North America are not. Dutch elm disease was first found in North America in 1930 in Cincinnati, Ohio. The evidence indicates that it was introduced on elm logs from Europe that were landed at ports on the eastern seaboard. The American elm had become an important amenity tree throughout the continent, being planted in urban sites to such an extent that it's no exaggeration to describe the plantings as urban forests. But by 1950, the disease was spreading through 17 states and into southeastern Canada. Today, Dutch elm disease occurs wherever American elms grow in North America. Countless millions of trees have been killed, with a corresponding multibillion dollar cost of removing and disposing of them and replacing the trees that were lost with new plantings. But the story has yet another twist. In May 1963 a shipment of American elm logs landed in the United Kingdom; this destroyed 25 million British elm trees.

There is no cure for Dutch elm disease, so the key to control of Dutch elm disease is sanitation. Dead, dying, or weak elm wood must be destroyed to eradicate both fungus and beetle. It goes beyond the aerial parts of the tree, however. The roots of adjacent elms tend to fuse together over time, resulting in a shared root system between several trees. This type of root grafting may occur between elms within 50 feet of one another.

The trees in our urban environment have rightly been described as an urban forest, which creates an environment that looks and feels right (top image). But Dutch elm disease attacks that forest. It can degrade our environment, turning a pleasant suburb (middle image) into a stark deserted landscape (bottom image).

When a single elm tree in such a group becomes infected, the fungus may move down the diseased tree into the roots and then into the next healthy tree through root grafts. The sanitation processes must include disruption of these root grafts by digging a trench two feet deep along a line around the beyond the longest branches of the diseased tree. Saving elms is a hard, costly job.

Fungi cause the majority of plant diseases, but they figure in only a minority of animal diseases. The fungus diseases of humans are called *mycoses*. The majority of mycoses, perhaps all, are caused not by dedicated pathogens, but by fungi common in other situations taking advantage of a host weakened in some way or of a particularly optimal set of environmental conditions. There are about 100 different human fungal pathogens, together with several other species that do not cause disease but prompt allergic reactions. The fact that there are relatively few mycoses does not mean that they are rare. What the human disease fungi lack in diversity, they possess in ubiquity. It must surely be true to say that almost everybody suffers from athlete's foot at some time in their lives. This is caused by a tropical import called *Trichophyton rubrum*. However, because this fungus so likes those nice warm and moist shoes we all wear, it is now distributed throughout the temperate climatic zones. Fortunately, athlete's foot may just be little more than a nuisance and can be successfully treated with over-the-counter remedies.

Another remarkable statistic about mycotic disease is that it is now unusual for a woman to go through her reproductive years without at least one significant infection by the yeast *Candida albicans*. This is a normal inhabitant of the human mouth, throat, colon, and reproductive organs. Usually it causes no disease but lives in ecological balance with other microorganisms of the digestive system. However, other factors such as diabetes, old age, pregnancy, and hormonal changes can cause *C. albicans* to grow in a manner that can't be controlled by the body's defense systems, and candidiasis results, with symptoms ranging from irritating to life-threatening.

We are prone to fungal invasion of the skin, nails, and hair. In addition to the aforementioned athlete's foot, ringworm is another fungal infection.

Ringworm is actually a family of mycoses; the cause is a fungus in one of two closely related genera, *Microsporum* or *Trichophyton*. Each fungus is very specific to a particular part of the body. Animals can also suffer ringworm diseases of skin and fur, and the fungi spread readily to humans.

We also breathe in a lot of fungal spores. Some of these may only challenge the immune system, but others find the lungs a good place to grow and can cause mycoses. *Cryptococcus neoformans* causes cryptococcosis; spores of the fungus are inhaled, begin to grow in the lungs, and can enter the bloodstream and be carried throughout the body. This fungus can cause cryptococcal meningitis, which is usually lethal. The fungus grows and sporulates well on pigeon droppings. This bird is so common that most of us must go near a source of infection fairly often. Histoplasmosis is another human respiratory disease associated with exposure to bird droppings. It is caused by the fungus *Histoplasma capsulatum*. Inhaled spores colonize the lungs, causing a disease similar to tuberculosis. Chicken farmers are the highest-risk group, but stay away from the roosts of large populations of starlings!

Aspergillosis is a respiratory mycosis caused by about eight different species of the green mold *Aspergillus*. Inhaled spores may cause asthmatic reactions but they may also grow within the lungs. The disease can be fatal if unchecked. *Aspergillus* species can produce enormous numbers of spores and are common molds (previously mentioned in chapter 1 as food contaminants that can produce toxins). The common name for aspergillosis, farmer's lung, suggests that the traditional victims are those who handle moldy hay and grain (farmers and brewery workers) but faulty or dirty air conditioners can cause spore-laden environments in modern office buildings.

Finally, the fungus *Coccidioides immitis* grows in desert areas throughout the Western Hemisphere and causes coccidiomycosis or valley fever. Inhalation of its spores infects the lungs. The young and healthy may suffer a slight cough, but it clears up. If the victim is not in the best of health, the infection can spread throughout the body and may be fatal.

Health statistics indicate the sinister fact that mortality from fungal disease is on a steady increase against a background of steady decline in mortality caused by all other infectious agents. There are a number of evident reasons for this. There has been an increase in diagnosis of fungal disease (that is, not necessarily more disease, but more of what does occur is being diagnosed as of fungal origin). To some extent this reflects the introduction of techniques, especially molecular methods, that can rapidly identify fungi. Secondly, increased availability of international travel has taken more people into the tropics, and tropical regions do seem to harbor more fungal pathogens. Thirdly, drug therapies used to manage the immune system in transplant and cancer patients have the unfortunate side effect of weakening the body's defenses against fungal pathogens. A fourth reason is that AIDS patients have similarly weakened immune defenses against fungi. Indeed, it is likely that most AIDS-related deaths are due to fungal diseases.

So what do we do about it? Across the board, whether plant or human diseases are concerned, the effort to counter the fungal menace must start with proper management: try to avoid giving the fungus any advantage. Effective management of plant diseases requires a good deal of information about fungal biology. Knowing the complete life cycle and infection conditions are important. Such knowledge may suggest control measures that target a vulnerable stage in the life cycle of the pathogen. The prime example of this is being the causal agent of black stem rust of wheat, which alternates between two hosts, wheat and barberry. Eradicating barberry plants growing in the vicinity of wheat breaks the cycle and protects the crop plant. Methods to control plant diseases aim at the pathogen itself, the development of a host more resistant to disease, and the attempt to use environmental features to reduce the effectiveness of the pathogen. The environmental features do not have to be terribly grand. Procedures like crop rotation, improving ventilation by wide spacing of plants, and arranging better drainage can do a lot to provide conditions less favorable for the growth of the fungus. Measures aimed against the pathogen include quarantine and sanitation measures. The first aims to avoid introduction of disease to an area where it is not established; the second aims at eradicating diseased plant material from the vicinity of the crop to reduce the inoculum. Ultimately, of course, we have to resort to application of chemical agents to injure or kill the pathogen.

In essence, control of fungal pathogens of humans is not so different. Avoid infection, don't encourage the environmental conditions that fungi prefer, and fall back on a panel of effective fungicides. Chemical control of fungi in human disease is more difficult than control of bacterial diseases because the chemistry of fungal cells is much more similar to that of human cells. A fungicide is toxic to the fungus, but it may also be toxic to the host and do too much damage to the healthy cells of the host. We need to understand the biological mechanisms that cause fungi to grow because this is fundamental to the development of treatments for fungal diseases in humans, other animals, and crops. With better knowledge of the fungal growth mechanisms, we can develop more specific and less environmentally damaging treatments for fungal diseases. Unfortunately, we are not doing the research that is necessary. People interested in fungal biology are being pensioned off from universities, research institutes, and commercial companies around the world at the very time when we most need their expertise. The worldwide market for treatments of human fungal diseases is worth approximately four billion U.S. dollars. It is divided into the three main sectors of over-the-counter remedies (the largest market sector), topical preparations (ointments and such that are put onto a particular part of the body), and systemic drugs (whole-body treatments given by mouth or injection). In 1995 the systemic sector accounted for sales of one and a half billion U.S. dollars and had a growth rate of 15 percent per year because of the accelerating use of these drugs to combat fungal infection resulting from immune-suppressing therapies used during cancer treatments and transplant management, as well as in HIV and AIDS patients. The most successful antifungal drug is Pfizer's Diflucan (fluconazole). It was first launched in 1988, and by 1997 sales had reached 900 million U.S. dollars.

On the other hand, the worldwide market for the use of fungicides in agriculture was estimated at almost six billion U.S. dollars in 1996, about 50 percent larger than the pharmaceutical market for antifungals. The two markets use similar chemical technology. Several products in both markets are derivatives of chemicals called azoles, which have two nitrogen atoms side by side. Azoles are relatively more toxic to the fungus rather than being nontoxic to the patient or host plant. There are not enough effective, broad spectrum antifungal drugs, and growing concern exists about the appearance of resistance to those antifungals that are presently available commercially. In particular, we have not yet found a nontoxic antifungal or fungicide. Unfortunately, we are unlikely to find such a drug because of the present lack of good scientific research in basic fungal biology.

Despite the colossal amounts of money invested in scientific research by industry, current corporate research strategies are aimed at corporate success and not the advancement of fundamental science. This is very clear in the current concentration on screening programs. Most pharmaceutical and agrochemical companies are involved in these. They represent safe science and are wonderful things for managers and accountants to understand because they have aims ("find something that works in this test") and measurable objectives ("we must screen 20 thousand compounds this year because our competitor is screening 15 thousand"). Science that is constrained by the profit motive is not good fundamental science because it is aimed at doing little, if anything, to advance our knowledge or understanding. Recognize it for what it is: it is good commercial science, aimed at providing sufficient marketable products to repay the megabucks that these programs absorb.

Even the much vaunted genetic engineering and cloning (genetic modification, or GM) programs are not necessarily good science either. True, they are technically wonderfully competent, and they provide tools that would enable us to unravel many mysteries. But when GM programs are directed toward corporate success, they have about as much to do with real science as the mechanic who repairs your PC has to do with designing the next generation of computer chips. Nevertheless, that's where the research money is being spent because the manager or accountant can easily understand that a very good way to make lots of money is provided by the cycle (1) be sole supplier of pesticide, (2) use GM to introduce the resistance gene into the crop and be the sole supplier of pesticide-resistant seeds, (3) make the crop plant sterile somehow so that fresh seed is needed every season. Even politicians can understand that. So they invest large sums in the close to market research to make it happen. It doesn't matter whether it is desirable (ethics is the responsibility of somebody else). It doesn't matter that it doesn't improve our overall understanding of biology (research scientists are being employed). The driving criterion is that marketable products will be produced. Markets will be satisfied. Some problems will be solved. But if only a tiny fraction of the funding devoted to this sort of market-lead science was used to fund research on organisms, we could advance understanding sufficiently to find alternative ways of achieving the same ends.

3

Decay and Degradation

Fungal Specialties

The biological character that sets off the fungi from the rest of the higher organisms, animals, and plants is that fungi release their digestive enzymes into their surroundings. They digest food sources outside themselves and then absorb the products of digestion as their nutrients. Almost any material can be digested as food by fungi. It could be argued that our most memorable encounters with fungi center around their abilities to deteriorate things. Fungi can cause decay and ultimate destruction of standing trees, fallen timber, and all sorts of timber constructions, which can be expensive! It was probably dry rot that made the *Speedwell* so unseaworthy that it could not accompany the *Mayflower* in crossing of the Atlantic Ocean toward the New World in 1620. And it was the decayed state of the *Investigator* that caused Captain Flinders, the explorer and navigator who first charted the entire Australian coastline, to put in at Mauritius and abandon the ship. England was then at war with France, and the French authorities on the island imprisoned him as a spy for almost seven years.

Wood is a prime target, but fungi have such an enormous range of digestive capabilities that they can cause deterioration of electrical and electronic equipment, leather goods, paper and textile products, and even optical equipment (some can extract mineral nutrients from glass!). They cause food spoilage, of course, and if food is destroyed in store, more than just financial problems result.

Even without digesting things, fungi can be destructive. Mushrooms may be soft, but they have been known to lift stone slabs and force their way through tarmac. Back in the 1860s, a famous mycologist called M. C. Cooke wrote *A Plain and Easy Account of British Fungi* in which he told of "a large kitchen hearthstone which was forced up from its bed by an under-growing fungus and had to be relaid two or three times, until at last it reposed in peace, the old bed having been removed to a depth of six inches and a new foundation laid." Cooke also tells of a comparable observation made by Dr. Carpenter: "Some years ago the [English] town of Basingstoke was paved; and not many months afterwards the pavement was observed to exhibit an unevenness which could not readily be accounted for. In a short time after, the mystery was explained, for some of the heaviest stones were completely lifted out of their beds by the growth of large toadstools beneath them. One of the stones measured twenty-two inches by twenty-one, and weighed eighty-three pounds . . ." However, not only nineteenth century building standards are prone to fungal attack. The Spring 1998 issue of the *Air France* magazine sported a beautiful color photograph of a fruit body of a fungus called *Coprinus* that had forced its way through the tarmac of a street in Paris. Another interesting pictorial example is a back cover picture on the April 1991 issue of the *Mycologist*, a magazine published by the British Mycological Society. This photograph shows fruit bodies of a puff-ball (*Scleroderma bovista*) coming up through a tennis court. The original hard porous court was made of fly ash and was overlaid in 1989 with 75 millimeters of gravel, and then a 20-millimeter layer of tarmacadam was rolled smooth. The fungal fruit bodies appeared in 1990. Reginald Buller, a Canadian researcher, did some experiments in the 1920s in which he put weights on the top of developing mushrooms to see how much pressure they could exert. He worked out that a single mushroom could apply a pressure of at least two-thirds of an atmosphere, which is about 10 pounds per square inch. It's all a matter of hydraulics, of course. The mushrooms can fill themselves with water and force their way through cracks and crevices. They are not doing it because of some perverse intention to break up pavement, but because in nature they need to push through soil and plant litter in order to bring their fruit bodies to a position from which they can release their spores to the winds. Evolution has equipped them with the tools that ensure that they will persist for generations.

Fungi can also decay timber very well. In the 200 years or so that Britain "ruled the waves" with its wooden ships, fungi took on the might of the Royal Navy and seemed to win quite often. Wooden ships were expected to last a long time. Today, it's rather remarkable when a battleship is taken out of mothballs to use its big guns to pound a coastline or two, but vessels did not become obsolete until about the middle of the nineteenth century. His Majesty's Ship *Royal William* lasted 100 years; it was built in 1719, took part in the relief of Gibraltar in 1782, and was the flagship of the Port Admiral at Spithead in 1805.

The Royal Navy always preferred oak for the hulls of its ships. If the native forests had been able to supply sufficient English oak, no other timber would have been used. Conservation legislation probably started with various timber preservation Acts of Parliament that controlled the use of oak and encouraged attempts at reforestation; these acts were intended to remove fear of British naval supremacy being lost due to oak becoming scarce through reckless felling. However, wooden ships have always suffered from rotting of their timbers. Cycles of alternate wetting and drying of parts of the woodwork, poor ventilation, and even the use of unseasoned wood in construction all favor the development of the fungi that have evolved alongside the trees to use the wood as a nutrient. Unseasoned

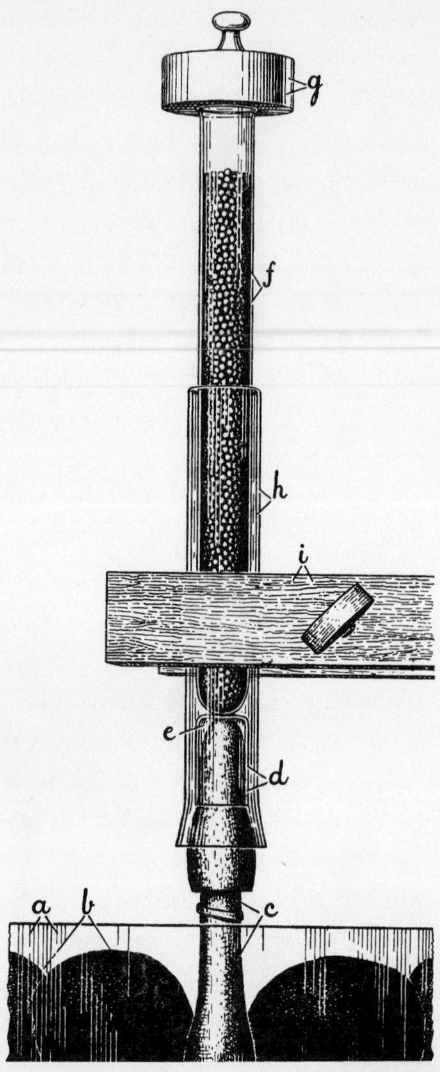

A *Coprinus* fruit body (c) that had been grown on some balls of horse dung (b) incubated in the laboratory. Reginald Buller put a glass tube (h) over the fruit body to stabilize it, then loaded the fruit body cap (e) with a test-tube containing lead shot (f) to a total weight of 150 grams. The fruit body grew quite happily, at a rate of 1 millimeter in 2 hours, with this weight on its head. Adding a further 50-gram weight (g) didn't reduce the rate of growth. Only when the overall loading was increased to 300 grams did the fruit body stem (c) bend and break. Three hundred grams is over half a pound, so that's not bad for a mushroom!

timber is bad news, or, as one of those old Acts of Parliament puts it: "'In buylding and repaireing Shippes with greene Tymber, Planck and Trennels it is apparent both by demonstration to the Shippes danger and by heate of the Houlde meeting with the greenesse and sappines thereof doth immediately putrefie the same and drawes that Shippe to the Dock agayne for reparation within the space of six or seaven yeares that would last twentie if it were seasoned as it ought." That is, do it right first time or the damn thing will be back in dock and you'll have to do it all over again!

Samuel Pepys gave, inevitably, the best and most forthright account of the problem. Not in his diary, but in a report to the Admiralty Board in 1684. As Inspector of the Navy he was required to carry out a survey of the fleet, especially 30 new ships. Unfortunately, his visit to Chatham docks showed only that: "The greatest part of these thirty ships (without having yet lookt out of Harbor) were let to sink into such Distress, through Decays contracted . . . , that several of them . . . , lye in danger of sinking at their very Moorings. . . . The planks were in many places perish'd to powder . . . and the ship's sides more disguised by patching . . . than has usually been seen upon the coming in of a Fleet after a Battle . . . Their Holds not clear'd nor aird, but (for want of Gratings and opening their Hatches and Scuttles) suffer'd to heat and moulder, till I have with my own Hands gather'd Toadstools growing in the most considerable of them, as big as my Fists."

The last two quotations contain all the information necessary for keeping timber constructions free of rot: use only properly seasoned timber and keep everything well ventilated. Unfortunately, their Lordships of the Admiralty did not learn their lesson. About 125 years after Samuel Pepys delivered his forthright criticism, the *Queen Charlotte* was built. This was a first rate battleship of 110 guns, launched in 1810, just five years after Nelson's victory at Trafalgar. The *Queen Charlotte* rotted so quickly that repairs in the first six years, even before it could be commissioned, cost £94,499 (more than the original cost of construction, £88,534!). By 1859 the total cost of repairs had risen to £287,837. She was broken up in 1892. At today's prices, the ship had cost the equivalent of about 2.5 billion U.S. dollars. That puts the cost of a B1 bomber in context!

If use of well-seasoned timber and provision of adequate ventilation are the keys to keeping your ship afloat, the same recipe will keep your house and home in safe condition. Structural timber is no different from other wood; it will decay unless kept dry. Proper building design is a key. Edgar Allen Poe understood well enough: "I scanned more narrowly the real aspect of the building. . . . Minute fungi overspread the whole exterior, hanging in a fine tangled web-work from the eaves. . . . In this there was much that reminded me of the specious totality of old wood-work which has rotted for long years in some neglected vault, with no disturbance from the breath of the external air." So, if the Ushers had taken Poe's pest control advice rather than dancing to his dramatic tune maybe the House would still be there.

The fungi that cause the rot are those common in woodlands, and for the most part they are indifferent to the carpenter's expertise. Wood that is always dry is immune to fungal attack. If used outdoors or in humid conditions indoors, all wooden structures eventually rot unless treated with some preservative. All kinds of wood are liable to attack; resistance to attack is relative. Soft woods are generally more susceptible than hard woods such as oak, yew, and teak.

There are three fungi that may be responsible for dry rot damage, but *Serpula lacrymans* is the chief culprit. When spores of this fungus fall onto damp wood, they germinate; this is one reason why wet timber is prone to attack. The threadlike hyphae that emerge from the spores penetrate into the wood and release enzymes that extract nutrients from the wood to support continued growth. The fungus can get nourishment only from wood (and wood-derivatives like paper and board), but the extraction of those food materials brings about changes in the chemistry and structure of the timber. The fungal hyphae may remain entirely within the wood with no external sign of their presence until severe rotting has developed; bulging and cracking are then the signs of attack, especially evident with painted wood. The fruit body of this fungus is not a mushroom. Rather, a flat, orange-brown or cinnamon-colored surface-hugging fruit body that ranges in size from a centimeter or so to a meter or more across may be formed. Each square centimeter is capable of producing

These photographs show some really bad news for the owner of the house! Dry rot in wall paneling (left photo) and roof beams (middle) of a house cellar. Fungi, like the dry rot fungus, *Serpula lacrymans,* can degrade almost everything that we make. The image on the right shows that it can even be worth protecting your computer data!

close to 1 million spores. So numerous are the spores that dusty deposits of them on furniture, floors, and other surfaces are often the first sign of dry rot noticed by the occupant of an infected house. But the spores are not the only way that *Serpula* can spread. Groups of hyphae join together along their length to form strands. Some of these can reach a thickness of 5 millimeters or more. The strands are invasive, and the cells of which they are made cooperate to grow away from the food source that is already infected to find other food sources.

The strands can translocate food materials efficiently, and this enables *Serpula* to spread over materials and structures from which it can derive no nutrition. When *Serpula* grows on wood it decomposes it and eventually reduces the wood to powder (that's why it's called dry rot). But when the chemicals that make up wood are digested, water is formed equivalent to half the dry weight of the wood. During active growth, therefore, the fungus can provide itself with the water it needs; although it must have moist wood to begin its attack, it can continue to grow into dry timber. Indeed, when growth is luxuriant, there may be excess water produced from wood decay, and this is exuded by the fungus in droplets. These are the tears suggested by the *lacrymans* part of the name.

In a real sense, the strands are explorers, and if wood is reached in a strand's wanderings, it is immediately attacked and eventually destroyed.

The strands are what make *Serpula* so dangerous. Strands can penetrate through the pores in bricks, cement, and stone, under tiles and other flooring, over plaster and other ceilings, and across anything that provides mechanical support. In the laboratory, strands have been grown across a full meter of totally dry plaster board, and they can do this as long as the originally infected wood continues to provide nutrition to the explorer. When the English country house Haddon Hall was being renovated, a large fruit body of *Serpula* was found in a stone oven; apart from a few strands it had no visible connection with anything else. The strands of the fungus passed through the joints of 9 yards of solid stonework, all the way back to a rotting floor elsewhere in the building. That was the first the owners knew about their rotting floor! *Serpula* strands can translocate food materials in both directions (in the laboratory, it can be shown that food can flow in both directions at the same time). So when the strand finds newly discovered timber to attack, maybe several yards away from the original, the whole infestation is integrated into a single organism that might become the size of the whole building!

A fungus called *Phellinus megaloporus* is found as frequently as *Serpula* in parts of Europe and was responsible for serious damage to the roof of the Palace of Versailles. This fungus requires very moist conditions and a relatively high temperature. It probably causes more rapid decay in oak than any other fungus, but it does not spread rapidly because it does not form strands. One of the other dry rot fungi, *Coniophora puteana*, is fairly common in buildings and characteristic of wood that is constantly wet. It does not extend its growth beyond the damp region. *Poria vaillantii* is not often found in buildings but was the scourge of damp mines, where it rapidly reduced timber roof props to uselessness. The wet and warm environments in many mines favor rapid growth of fungi. Mine timbers 5 to 6 inches in diameter may rot completely in less than a year, and while rotting the wood, the fungus produces long threads of white mycelium hanging down from the roof timbers.

Wooden ships, buildings, and mine shafts are not the only ways we have of feeding fungi. Soon after railroads began to spread across the world's continents, it was realized that the wooden cross ties (sleepers), being half buried in the ground, decay rapidly. In the United States, *Lentinus lepideus,* a particularly troublesome mushroom, became known as the “train wrecker”. It grew even on ties made from evergreen timber treated with creosote, and some train derailments were blamed on the decomposition of cross ties by this fungus. The U.S. Department of Agriculture was so concerned that for several years it banned the import of the related shiitake mushroom for fear of the damage it might also cause. Because shiitake is not especially destructive, the USDA has since relented.

Not all wood deterioration results in destruction of the timber. There are other fungi (sap-stain fungi) that cause serious losses to the lumber industry by discoloring wood. Although the timber is not weakened, the discoloration renders the wood unfit for most purposes, so its value is downgraded. The sap-stain fungi are divided into two groups on the basis of whether they penetrate the wood. Where the fungus growth is superficial, the stain does not penetrate the wood and may be removed by planing; but when the fungus penetrates into the wood the blemishes are too deep-seated to be removed. The value of stained lumber is reduced because it cannot be used with a natural finish. Staining develops while the lumber is being stored. Sap-stain fungi develop most rapidly when the wood has a high moisture content and the weather is warm. So the often repeated preventatives apply: use well-seasoned (i.e., dried) timber and keep it well ventilated. Some staining is actually desired for special cabinet work. The fungus *Chlorociboria aeruginascens* produces a characteristic bright blue-green color in oak and other deciduous trees, and wood stained in this way has been used for ornament, usually marquetry, in products called Tunbridge ware. There is even a British patent issued in the early years of the twentieth century covering the artificial infection of trees with *C. aeruginascens* to generate stained timber. Woodturners appreciate the pleasing color patterns that may result from discoloration in wood caused by the early stages of rotting. Dark lines are often found in infested wood. They are due to interactions between different fungi growing in the timber and the effect of fungus growth on the chemical structure of the wood. It is called spalted wood, and the apparently random lines create natural decorative designs for bowls and vases. If rotting has gone too far, the wood is unworkable, so finding a log in the right condition can be a challenge.

Attempts at wood preservation date back a long time. The most authoritative instructions were those given to Noah: "Make thee an ark of gopher wood; rooms shalt thou make in the ark; and shalt pitch it within and without with pitch." Other ancient texts record the preservative properties of the oils expressed from olive, cedar, larch, juniper, valerian, and so on. The Romans knew well that, ironically, wood kept continually wet is less liable to rot. According to Pliny: "The pine, the pitch-tree, and the alder are employed for making hollow pipes for the conveyance of water, and when buried in the earth will last for many years." And according to Christopher Wren: "Venice and Amsterdam being both founded on wooden piles immersed in water, would fall if the constancy of the situation of those piles in the same element and temperature did not prevent the timber from rotting." The fact is that fungal decay of wood requires moisture and air. While the timber is submerged it is relatively safe, perhaps for hundreds of years, but if excavated without concern to its preservation, it can decay in a few months. Various treatments to prevent rot have emerged over the years. The oils of coal tar, obtained by distilling coal, which are heavier than water, are known as creosote. Although introduced in 1838, creosote is still probably the most successful of all wood preservatives. There are also other preservative oils in use, as well as water-soluble chemicals and chemicals dissolved in oils and nonaqueous solvents.

As for treatment, the traditional eradication strategy is removal and replacement of infected timber. This inevitably involves major building work, use of considerable amounts of chemicals, and a great amount of inconvenience and expense. A radical alternative approach emerged in Denmark in the early 1990s. It was observed that rafters directly under a black roof were rarely affected by the dry rot fungus. The temperature in such locations was often quite high, and the moisture content in the timber was low. Then laboratory experiments showed that *S. lacrymans* is very sensitive to high temperatures, and the idea emerged to use heat treatment to cure dry rot. Heat treatment, that is, of the whole building, or at least those parts of it most affected. Once a building surveyor has mapped out the extent of the attack and the details of the construction, sources of moisture are identified and eliminated. Loss of strength in load-bearing timbers must be evaluated because there is no alternative to their replacement if they have been weakened too much. The final preparative step is to simulate the treatment on a computer to optimize energy usage. Temperature sensors are put in places that are difficult to heat, for example, in the center of walls and beams. Then the building is covered with a 100-millimeter thickness of insulation material on scaffolding. Finally, warm air from oil or gas burners is blown into and around the insulated zone. The temperature rise is monitored twice each day until it reaches 40°C, at which it is held for at least 24 hours. The laboratory tests indicate that this is sufficient to kill the fungus in the wood. When the heat treatment is finished, timber repairs or replacements identified in the initial survey can be done, and the building can be restored to normal use, although continued vigilance against reinfection is necessary. The cost of a heat treatment is 10 to 50 percent less than a standard repair, and so far results are good enough to encourage the technologists to look for alternative ways of raising the temperature of a building, such as microwaves.

Fungi are quite capable of growing on waxes, paints, leather goods, and all forms of textiles, from the finest cotton to the heaviest canvas. Possibly the most exotic example of fungal deterioration is that which occurs, especially in tropical regions, on optical equipment such as binoculars and telescopes. Although glass alone probably does not support the growth of fungi, organic material on glass lenses or prisms, even in minute quantities (flakes of skin, dead mites, fingerprints), can provide sufficient substrate for fungus growth, which may then spread over the glass surface. In the majority of cases, this sort of growth can be simply wiped from the glass to clean it, but sometimes the glass is definitely etched by the fungus and a fine network of unwanted scratches are left behind on the lens or mirror. Some laboratory experiments indicate that certain soil fungi will grow on silica gels, dissolving other minerals they need from the silica and extracting nitrogen and carbon from volatile gases in the atmosphere.

Faced with a catalog of destructive potential such as previously discussed, it's difficult to avoid the conclusion that there is some purposeful malice at work, aimed at destruction of human habitation, transport, and possessions. Of course, that's too fanciful to be true. The truth is that fungi have been evolving for many hundreds of millions of years to live in natural wilderness habitats. Those habitats include fallen timber and all sorts of plant and animal debris, leaves, fur, and feathers. But it's not a uniform mixture. There may be a large piece of wood here and another piece over there with all sorts of dirt, soil, and rocks in between. The evolutionary strategy that works for the most efficient wood decay fungi is to grow on one piece of wood and use some of the nutrients to send out exploring filaments to grow across the soil and rock to find the next piece of wood. That's what the forest floor is like for a fungus, islands of nutrients scattered over a mineral desert. Now, imagine that someone comes along and builds a country cottage: timber beams, stone floors, wooden stairs, and brick walls, with books, shoes, fabrics, and fast foods scattered around. Any passing fungal spore is going to see islands of nutrients scattered over a mineral desert, especially if this person keeps the windows closed and the cottage is nice and moist.

An organism becomes a pest when it does what it normally does, but in the wrong place. A fungus that rots the wood of your roof timbers is a pest; the same fungus in the forest is doing an essential service for you. Fungi that decay organic material are a benefit to humanity in three very important ways. First, organic debris is continuously being removed from the environment. Second, because the debris is removed by being digested, large quantities of carbon dioxide are released to the atmosphere and made available again for use by green plants in photosynthesis. Third, the matter that's left after the decay has proceeded as far as possible in the short term is the humus that forms the very structure of soil.

It's been calculated that a single large broad-leaved tree has at least 1 million leaves. At the end of the growing season, the dead leaves fall to the ground. They'll weigh about 200 kilograms. In five years, that's a tonne (metric ton) of dead leaves from one tree. I don't know if anyone has ever counted how many deciduous trees there are in the world, but my guess is about 1 billion or so. And every five years they make 1 billion tons of dead leaves. If fungi and bacteria were not digesting those leaves, we'd have an awful lot of leaves to wade through to get to the office, and furthermore, the atmosphere would rapidly run out of carbon dioxide. Each growing season, tremendous quantities of atmospheric carbon dioxide are used by green plants in photosynthesis. As a result, the carbon is chemically bound, first in sugar and eventually in other carbohydrates, proteins, fats, and in all of the compounds that make up living things. While they are binding the carbon, the plants break up water and release the oxygen into the atmosphere. So, continued photosynthesis is necessary to maintain atmospheric oxygen, and carbon dioxide is necessary for continued photosynthesis. We depend on those natural digesters of organic debris for continuous recirculation of carbon by returning of carbon dioxide to the atmosphere. One estimate is that about half the atmospheric carbon dioxide is bound organically every year. If decay of wastes should cease, we would not have to worry about the greenhouse effect because without carbon dioxide, photosynthesis would cease and there would be no oxygen to breathe.

Balance between production and decay is crucial. During the Carboniferous period, the rate of formation of organic material exceeded the rate of decay, and an amount of material corresponding to 4 times the total carbon dioxide content of our present atmosphere accumulated. This covered a time interval from 360 million to 290 million years ago. The name *Carboniferous* originated in the United Kingdom, where it was first applied (in 1822) to the coal-bearing strata of England and Wales. Around 330 million years ago, tropical forests and swamps covered large areas of what would eventually become eastern North America and northern Europe. These areas had warm and humid climates because at that time they were situated in the tropics, immediately north of the equator. Such conditions promoted lush growth of the vegetation and marine organisms. These areas became our present day suppliers of coal (and oil and gas). If the balance were to be disturbed to that extent again, we'd be up to our ears in dead trees.

It's not just a case of buildings and possessions being damaged by fungi, there's the case of food spoilage as well. If you think about an average food, say an egg salad sandwich on rye bread, you will appreciate that foods are complex and very varied ecosystems. That sandwich I've just mentioned features highly localized regions of concentrated carbohydrates, concentrated proteins, and concentrated fats. These regions differ drastically in chemical and physical characteristics and are likely to have been changed by the various types of processing to which the different components have been subjected. Inevitably, therefore, there is a wide range of organisms that may cause food spoilage. They tend to have in common the feature that they can tolerate environmental extremes. Most foods are pretty dry. They are either genuinely dry, like bread, corn flakes, and flour, or physiologically dry like molasses and brine, in which the high concentration of sugar or salt make the water in the fluid difficult to access. Fungi that contaminate these sorts of food are highly specialized to dry conditions. Other extremes that food spoilage fungi might tolerate are acidic conditions (in pickled foodstuffs, for instance), and they may even survive the high temperatures used in pasteurization. Finally, some can tolerate the preservatives that are added to food. Indeed, some off-flavors and off-odors, especially those that resemble petroleum products, are caused by the contaminating fungus metabolizing the preservative. Food spoilage by fungi is sometimes a matter of public safety; remember those mycotoxins that some fungi produce? However, this is relatively rare. The main issue is more likely to be consumer acceptability. Fungal food spoilage is usually fairly obvious, either to the eye (moldy bread, for instance) or to the nose. The sense of smell is still the most widely used, and the most sensitive detector for off-odors; the nose detects volatile compounds produced by the fungus during growth. The smell of a musty or moldy grain shipment will determine its value and acceptability after transport or storage. So spoiled products have a real image problem. Even if the spoilage is actually benign, the look and smell will make the food unacceptable. The result is that food preparation and processing industries cannot take risks. Contamination rates may be extremely low; maybe something much less than a 1 percent contamination frequency in a production run of 100,000 cans of a fruit drink, for example. But the risk of consumer resistance to the possibly contaminated product is too great, and even at that low level of contamination, the product should be recalled and destroyed. Detection of spoilage organisms sufficiently early in the food preparation to enable them to be removed without damage to the product is the key to control. Much effort is being devoted to research aimed at developing electronic artificial noses, or to give them their politically correct name, volatile-compound-mapping machines. However, trained people in sensory panels still remain the principle analytical safeguard against contamination by food spoilage fungi. And the use of preservatives in the food products themselves is our ongoing defense against the fungi. Continuous development of new approaches is needed, however, because the fungi continue to change. Common contaminants may mutate at any time to become tolerant of preservatives or more troublesome in the processing train. In addition, as commerce becomes more international, entirely new food spoilage fungi can be introduced to an industry from some other part of the world. The spoilage fungi are just as international as the industry they attack.

4

Joining Forces

Fungal Cooperative Ventures

Fungi, animals, and plants have coexisted since these three groups of higher organisms originated. Living close together for a long time can cause neighbors to be at each other's throats or in each other's pockets. Different fungi have followed both of these routes. We have already seen how the "at each other's throats" metaphor emerges with fungal diseases and toxins that enable fungi to have some advantage over animal and plant adversaries. But there are several fungi that have taken the opposite route by treating plants and animals as partners in mutually beneficial relationships. Scientists call such arrangements *symbiosis* or *mutualistic associations.* The organisms concerned (often two but sometimes three) live in such close proximity to each other that their cells may intermingle and may even contribute to the formation of joint tissues. In these associations the partners each gain something from the partnership and are more successful than they would be without the association. Fungi are involved in some very ancient mutualistic associations. Lichens, basically combinations of fungi and algae, can be found in some of the most inhospitable environments. Another association is a mycorrhiza, which is a fungal infection of plant roots. At least 95 percent of land plants rely on them, especially for mineral nutrients, and in return the fungus gets some sugar formed by the plant in photosynthesis. There's more to it than that, though. In high mountain areas when conditions turn stressful, the fungus supplies the plant with sugars and other nutrients it gets from its abilities to degrade waste materials in the soil. By sharing the products of these activities with its plant host, it keeps its host alive until the sun shines again. In lush lowland forests the soil is so full of mycorrhizal fungi that they connect the trees together, so trees and their seedlings can exchange food and messages. Leaf-cutting ants chew up leaves to make compost to grow one particular fungus that produces a nutrient fluid the ant grubs need to drink. A couch-potato of a fungus that sits there while the ants charge around the forest collecting leaves for it: who's using who?

Lichens form a large and very widespread group of organisms that are associations between fungi and photosynthetic organisms. The most ancient associations, in evolutionary terms, are between fungi and blue-green algae. Blue-green algae are better called cyanobacteria because they are bacteria rather than algae, but they do have chlorophyll and can make their own food by photosynthesis. Cyanobacteria were the first organisms to release oxygen into the Earth's atmosphere, so probably 3 billion years ago they started the evolution from which the atmosphere we rely on today emerged. Their chemical activities in the shallow seas of those ancient days precipitated insoluble calcium salts into the shapes of their colonies, and these (now called stromatolites) became the first fossils on the planet.

As the more advanced plants, specifically algae, evolved, the fungi (mostly relatives of cup fungi, or ascomycetes) developed the partnerships that are now known as lichens. Lichens have always been thought of as classic examples of symbiosis; because of their unique partnership, they are able to live in places that are inaccessible to other organisms. The fungus takes in water and uses its externalized enzymes to extract nutrients from the soil and even from the rocks themselves. The algal partner uses photosynthesis to provide carbohydrates to the partnership. However, more recent studies question just how truly mutually beneficial the relationship might be. Most of the body of the lichen is fungus; the algal partner constitutes only 5 to 10 percent of the total biomass. Also, although some of the algal partners can grow alone without the fungus, most of the fungi can live only in the association with the alga. Indeed, microscopic examination shows that fungal cells actually penetrate the algal cells in a way similar to pathogenic fungi. The belief currently is that the fungus in the lichen is really parasitizing the alga and using the products of algal photosynthesis to feed the fungus.

Whether it is a mutually beneficial association or a case of the fungi taking algal hostages, there are about 20,000 species of these unique entities in the living world. They vary in size, shape, and color. Some are flat and firmly attached to the surfaces they grow on (like those yellowy-brown disks that are scattered over building stones in country-clean areas. But others are scaly, leafy, bushy, or hang in strands from their supports. For reproduction, several lichens produce flakes (called *soredia*) as a powdery mass on the upper surface of the lichen or fragile upright columns (called *isidia*). Either way the fragments include both fungus and alga and are easily dislodged to be blown about in the breeze, like spores. They can begin a new colony if they land in a suitable place.

Lichens have a remarkable ability to thrive where no other organisms can exist. They can tolerate temperature extremes from the heat of deserts to the cold of Arctic and Antarctic wastes. They survive drought by extracting moisture from mists and fog. Understandably, lichen growth can be on the slow side. In the Arctic, lichen growth is probably 2 inches in 1,000 years, but in less extreme environments (like on the roof of your country cottage), they may grow as much as an inch or two in 10 years! Some large colonies have been estimated to be around 5,000 years old. The dependence of lichen on the atmosphere and rainfall makes them highly sensitive to atmospheric pollution. Acidic rain and sulfur dioxide kill many lichens, so cities in industrialized countries may have few lichens because of the poor air quality. If you do have lichens growing on your stonework it's probably good for your lungs!

The nutritional value of lichens is similar to cereal seeds, though they do not make major contributions to human food. In the harsh areas where lichens grow, native peoples use them as food supplements. One lichen, which occurs in the deserts of the Middle East, may have been the manna that fell from heaven to rescue the children of Israel from starvation in the Old Testament story: ". . . and in the morning the dew lay round about the host. And when the dew that lay was gone up, behold upon the face of the wilderness there lay a small round thing, as small as the hoar frost on the ground. And when the children of Israel saw it they said one to another, It is manna: for they wist not what it was. And Moses said unto them, This is the bread which the Lord hath given you to eat." (Exodus 16: 13–15).

Lichens are much more important as animal food, especially in the Arctic, where lichens can form as much as 95 percent of the diet of reindeer. In less harsh conditions, most mammals of these wildernesses will supplement their normal winter diets with lichens. And sheep in Libya graze a lichen that grows on the desert rocks, grinding their teeth away as they chew it off! The extreme life style of lichens leads them to produce several exotic chemicals, some of which may be useful to humans, such as antibiotics, essential oils for perfumery, and dyes for textiles. And there are probably many others awaiting discovery and exploitation.

The lichen partnership contributed to the earliest steps in evolution of the Earth's environment, but the partnership that contributed by far the most must be the mycorrhizal association between fungi and the higher plants. In this relationship the roots of the plant are infected by a fungus. But the rest of the fungus continues to grow through the soil, digesting and absorbing nutrients and water and sharing these with the plant. This was discovered by a German botanist called Frank in 1885. He revealed this curious relationship between fungi and the smaller roots of higher plants. He claimed that it resulted in a compound structure composed of both plant-root tissue and fungus mycelium and called the compound structure a Pilzwurzel, or fungus-root. The name has now been translated into a mixture of Greek and Latin to become *mycorrhiza* (which still means fungus-root). Frank suggested that the mycorrhiza might be of fundamental importance in the nutrition of trees. Later studies have shown that just about all of Frank's interpretations were correct. Mycorrhizas are indeed, discrete organs similar to lichens in that they are composed of two partners: a fungus and a plant root.

It seems that plants roots are just not up to the job of supplying the plant with everything it needs. Certainly, all plant roots are much bigger and more unwieldy than fungal hyphae, and the root-hairs occur only at the apex of the finest roots. There are far more mycorrhizas. So the mycorrhizas enable the plant to connect with a vast network of fine hair-like fungal cells, which like fungi everywhere, are exploring and seeking out fresh nutrients. The external mycorrhizal network is made up of such a large number of thin cells that it has an enormous surface area and equally enormous capacity to absorb things from the surroundings. These are the advantages over the root alone: active exploration and a large surface area for absorption. In addition, external parts of a mycorrhiza form an interconnected network through which nutrients are readily transported, which creates an ideal nutrient search, capture, and delivery device for the plant to use. But this is a mutualistic relationship, so the plant pays for the privilege of using this fungal device by sharing up to twenty-five percent of the products of its photosynthesis with the fungus. Despite this tax on its activities, the plant grows better than it would without the mycorrhiza. Each partner benefits from the special abilities of the other, and the two together make out better than either would if left to grow on its own.

Most terrestrial plants have mycorrhizal relationships. The arrangement has become the rule rather than an exception. And although most fungi are not mycorrhizal, there are common and important examples from all the major types of fungi. The mycorrhizal fungi we are most likely to meet are the mushrooms we see in wooded areas. The aforementioned *Amanita* and *Boletus* are mycorrhizal partners with trees and other forest plants, as are chanterelles and truffles (although truffles are not mushrooms).

Some mycorrhizal fungi form a mat of fungal tissue around the root, the fungal cells penetrating between the cells of the plant root but never actually crossing the plant walls. These are called ectomycorrhizas and are common on conifer trees as well as on some deciduous plants. In another mycorrhizal partnership (called endomycorrhizas), the fungal cells penetrate plant cell walls. Inside the plant cells, the fungal cells branch out to make structures that absorb materials from the plant cells. These endomycorrhizas are made by the most primitive fungi and can even be identified in the most ancient of plant fossils. Therefore, these could be the first mycorrhizas, which helped the first plants to invade the land. Today they occur on many deciduous trees, shrubs, and crop plants. By greatly increasing the absorbing surface of a host plant's root system, mycorrhizas improve the plant's ability to withstand drought and other extremes, like temperatures and acidity. A mycorrhizal plant also has an improved supply of mineral nutrients and some protection from pathogens in the soil. The importance of mycorrhizas is most clearly seen in new plantings. Pines are completely dependent on mycorrhizal infection for normal development. When pine is planted in recently cleared ground where the tree has not grown before, it is necessary to supply suitable mycorrhizal fungi. The ground can be inoculated with roots, with humus obtained from established pines, or with pure cultures of the proper fungus produced in the laboratory. When this is done, vigorous growth is seen in seedlings placed nearest to the point where inoculation was made. Infection spreads rapidly from plant to plant after the mycorrhizal fungus is established in the soil. Healthy plants have mycorrhizas; where these structures are lacking or poorly developed, the plants will be stunted. Seedlings with mycorrhizas weigh more, have a smaller root–shoot ratio, and contain greater quantities of nitrogen, phosphorus, and potassium than seedlings that lack mycorrhizas. In stressful sites, tree seedlings that are not infected show symptoms of nitrogen and phosphorus deficiency and eventually die. Every afforestation project is doomed to fail unless a suitable mycorrhizal fungus is introduced along with the trees. A high degree of specificity exists in the establishment of the mycorrhizal relationship in some cases, to the extent that the distribution of certain plants might be governed in part by the distribution of fungi capable of forming mycorrhizas with them. In other cases there is little or no specificity. At the extreme end of the mycorrhizal spectrum are associations between fungi and plants that lack chlorophyll. In these nongreen plants (of which the Indian pipe, *Monotropa,* is an example) the entire root system is involved in a mycorrhizal relationship. Since the plant is unable to photosynthesize, it is completely dependent on the fungus associated with its roots for supplies of all nutrients, including all of its energy-containing carbon compounds derived from the fungal degradation of substrates in the soil. In this extreme case the plant gives little, if anything, to the partnership and is actually a parasite on the fungus infesting its root system.

Aside from pathogens or mycorrhizas some, maybe many, plants harbor other fungi that affect their growth. These fungi are called *endophytes* because they exist within their host plants. Of course, parasites also do this, but parasites damage the host, whereas endophytes are at least harmless and may be beneficial. They may be harmless or beneficial to their host, but their benign nature is not universal. Endophytes became a hot research topic when it was found that some which live entirely within grasses are responsible for the toxicity of grasses to livestock. The particular grass concerned is called tall fescue. This is a hardy, pest-free, and drought-tolerant grass that was introduced to the United States in the early 1900s as a forage grass from Europe. The U.S. plantings contained an endophyte, however, that produced toxins related to those of the ergot fungus. So animals fed on tall fescue suffered illnesses similar to the ergotism of the Middle Ages. Pregnancies were disturbed, so foals and calves were lost, and some cattle lost limbs and suffered gangrene. And all were more sensitive to other diseases because of a general reduction in vitality. The ergot alkaloids the endophyte produces in tall fescue are bad for the livestock that graze on the plant, but they improve the health of the plant by protecting against attack by grazing insects, not to mention cattle and horses! The endophyte-infected tall fescue must be avoided in livestock feed, but it is actively encouraged in fescue seeds intended for amenity sites, especially golf courses. Because the infected grass produces its own pesticide, use of chemical pesticides in those locations can be reduced significantly.

There are numerous other endophytic fungi. A functional relation to the host is not always obvious, and some may be simple passengers, living in the inner spaces of the plant in much the same way as they would live in any other moist, secluded place. But it seems likely that some intriguing stories await revelation, such as the endophyte in oak leaves that remains dormant until a sedentary insect activates it by chewing on the leaf. The fungus responds to the insect attack by becoming a pathogen, killing a zone of the leaf surrounding the insect so that the insect dies for lack of live leaf tissue to feed on. With the insect pest eliminated, the fungus returns to being harmless, and the oak's leaves can photosynthesize in peace!

Fungi are not nasty to all insects, however. Fungi produce chemicals to attract insects to particular smells and tastes so that the insects can carry fungal spores around. On the other hand, insects feed on fungal fruit-bodies like mushrooms. But there are a few other examples of insects and fungi evolving together to become even more intimately involved.

The first example is the leaf-cutter ant, which cultivates fungi in its nest as an ongoing food supply. There are several sorts of ants in Central and South America that are known as leaf-cutter ants. The workers cut pieces from leaves on trees and carry them back to the nest. Because they usually carry the leaves in their mandibles so that the leaf extends over the ant's head, they are also called parasol ants. The caste that produces the largest animals among the several million ants in an average nest is the soldier; a 20-millimeter-long ant that is responsible for protecting the colony and its trails against intruders. The most numerous caste in the colony is the worker caste (these ants are about 8-millimeters long) that forage in the forest in search of leaves. They can cut leaf pieces bigger than themselves and then carry them back to the colony. Once delivered to the nest, smaller workers about half the size of the foragers chew the leaves into smaller pieces and carry it into brood chambers. Then the smallest ants, only 1.5-millimeters long, take over. These are the cultivators of the fungus garden. They clean the leaf pieces and then inoculate them with fungus mycelium taken from the existing garden. The cultivators then continue to maintain the fungus garden, but they also tend the larvae. The fungus cultivated by leaf-cutter ants does not produce spores, but it does have special cells that exude a sort of honey-like dew that the cultivator ants collect and feed to the larvae. This fungus is related to the mushroom-producing fungi, but it has never been found living free in the forest; the leaf-cutter-ant fungus is always associated with leaf-cutter-ant nests. This is a mutual dependence. Although the mature ants can get nutrients from the plant sap and tissue of the leaves (so that the fungus is merely a diet supplement), the larvae depend entirely on the fungus to digest the leaves and supplement the protein and vitamin content of the leaf material.

Since the fungus does not have an existence separate from the ants, it must be carried from the parental nest by newly mated females as part of the mating flight. The new queen mixes the fungus inoculum she carries with some suitable plant material and lays eggs on it as soon as the fungus begins to grow. She lays about 50 eggs each day. The first to hatch become workers who eventually establish a nest with a thousand or so interconnected chambers that might be excavated 5 meters down into the forest soil and be able to house a colony of 5 to 7 million individuals. The demand for leaf material as the colony grows to this sort of size is enormous. In the tropical rain forests of Central and South America, leaf-cutter ants are the dominant herbivores. That "dominant" includes the humans of the forest. Leaf-cutting ants compete successfully with humans for plant material and are therefore counted as important pests. Potential losses each year (assuming no control measures are used) could exceed one billion U.S. dollars. Around 50 agricultural and horticultural crops and about half that number of pasture plants are attacked. None of this is new, of course. In the last quarter of the nineteenth century, leaf-cutting ants were described as "one of the greatest scourges of tropical America." It's been calculated that in a tropical rain forest leaf-cutting ants harvest 17 percent of total leaf production. Nests located in pastures can reduce the number of head of cattle the pasture can carry by 10 to 30 percent. Statistics like these reveal how leaf-cutting ants can become dominant exploiters of living vegetation, amply justifying the description *dominant herbivore*. The combination of a top-of-the-range social insect with a top-of-the-range fungal plant-litter degrader seems to be the key to this success. The social insect has the organizational ability to collect food material from a wide radius around its nest, but the extremely versatile biodegradation capabilities of the fungus enable the insect to collect just about anything that's available. The total number of species of trees per hectare in most plant communities increases from the poles to the equator. For example, the coniferous forest in Northern Canada has 1 to 5 species per hectare, the deciduous forest in North America has 10 to 30 species per hectare, and the tropical rain forest in South America has 40 to 100 species per hectare. The tropical rain forest has enormous chemical and physical diversity in its plants, and this presents a major problem to most herbivores. Most plant eaters have a narrow diet tolerance because evolution has equipped them with only a limited range of digestive enzymes. Plant-eating insects usually only eat one plant. The leaf-cutting ants of the tropical rain forest have, on the other hand, a very wide breadth of diet. These ant colonies are able to harvest 50 to 80 percent of the plant species around their nests. This is almost entirely due to the broad range of degradative abilities of the fungus they cultivate. Some plants do protect themselves by producing deterrents that inhibit cutting, pick-up, or feeding. These include toughness, production of sticky latex, and a wide range of defensive chemicals. These natural defenses have recently been subject to intense study in the attempt to discover ways in which the insects might be controlled in the field. There's a certain irony in the fact that the tropical rain forest is lush and green because all those mycorrhizal fungi in the roots of the trees gives the plant that extra something that enables it to grow with tropical exuberance. And then along comes a six-legged army of harvesters to cut down all those lush green leaves to feed another fungus. Yet maybe it's not as simple as irony. Maybe we're missing the plot here!

Leaf-cutting ants cultivate their fungus in South America, but in Africa, the insect partner in a similar relationship is a termite. Termites are responsible for the bulk of the wood degradation in the tropics. Most of them carry populations of protozoans in their guts to digest the plant material and release its nutrients. Termites in the family Microtermitinae have evolved a different strategy. They eat the plant material to get what nutrition they can from it and then use their feces as a compost on which they cultivate a fungus from the mushroom genus *Termitomyces*. The wider range of enzymes the fungus can produce digests the more resistant woody plant materials, and the fungus becomes a food for the termites.

Some fungus-cultivating termites build mounds consisting of fecal compost and fungus above their underground nests. Different termites produce mounds of different size and shape. Chimney-like termite mounds up to 30-feet tall are common in several parts of the bush in Africa. Inside, the mounds have many chambers and air shafts that ventilate both nest and fungus culture. All termite larval stages and most adults eat the fungus. The termite king, queen, and soldiers are exceptions, being fed on salivary secretions exuded by the workers. The fungus cultivated by the termites is one that produces mushroom fruit bodies, but not in an active mound. In some way the termites prevent the mushrooms from forming in the mound. In the rainy season the termites may take portions of the culture out of the mound to fruit on the ground nearby. Abandoned mounds also produce mushrooms after the termites have left, and these are some of the largest mushrooms you can find, being up to about 1 meter across the cap.

Fungus-growing termites are pests because they attack wooden structures. By eating through the wood, they leave a maze of galleries that destroy the strength of the timber. Insecticides and fungicides can help to control this pest.

The final example of an intimate interdependent association between an insect and a fungus concerns a wood-boring beetle in the family Scolytidae. These are usually found inhabiting the trunks of living trees that have been under some sort of stress (drought, air pollution, etc.). They may also be found in trees that have been recently cut or blown down. The female beetle bores into the tree, lays eggs on the tunnel wall, and inoculates the wood with fungal material she has carried from a previous nest. The beetle has a specialized body cavity (called a mycetangia) in which the fungal spores and mycelium is transported. By the time the eggs hatch, the fungus will have grown over the tunnel walls, using its enzymes to digest constituents of the wood. This fungus "lawn" (called, rather fancifully, *ambrosia*) provides the developing young larvae with a readily digested "food of the gods." Eventually, the larvae pupate and subsequently emerge as adults with a supply of fungus in their mycetangia. Because they are "gratuitous" food for the larvae, the fungi have become known as ambrosia fungi and the insects as ambrosia beetles.

5

Fungi in Medicine

Antibiotics and Other Pharmaceuticals

We probably have all heard the story about penicillin being discovered by chance when some research egghead named Fleming returned from holiday to find that his bacteria had been slaughtered by a rampaging fungus. Different versions of the story differ as to whether there and then he recognized the thing as the first antibiotic. And they also differ in the way they describe if and how Britain and the United States collaborated to produce the material that would save humanity from disease. Well, no certain version of the story can be either entirely true or entirely wrong. But I don't intend to contribute to the variety of stories, because I don't want to start here. Rather, I want to start by painting a picture of the magnitude of the change in lifestyle that was brought about by antibiotics. Today we take antibiotics for granted. When you have a mildly sore throat, you take an antibiotic; when you have a slight injury, you take an antibiotic, "in case of complication." It's all become so trivial and ordinary. But the changes created by the immediate availability of antibiotics were far from trivial. For the first time in human history, they brought freedom from fear of inevitable death from blood poisoning of various sorts. A death waiting for everyone for the most trifling of reasons.

In 1940 the Second World War was really beginning to get into gear. In the summer of 1940, Hitler's forces dominated Europe from the North Cape (a promontory on a northern Norwegian island, which is the most northerly point of Europe) to the Pyrenees. Winston Churchill succeeded Chamberlain as British Prime Minister on May 10, and some of the darkest days of World War II followed: Dunkerque, the fall of France, and the blitz. The British army had left most of its weapons on the beaches at Dunkerque, but the United States, after the fall of France, began the first peacetime conscription in its history. U-boats of the German Navy initiated the Battle of the Atlantic in June 1940, a conflict that would not be resolved until late 1943. The Battle of Britain was fought in the air from August 1940, when the German Air Force launched daylight raids against ports and airfields. But the new device, radar, so much increased the effectiveness of Royal Air Force fighters that the magnitude of German losses forced them to switch to night bombing at the end of September 1940. Between then and May 1941, there were 71 major air-raids on London and another 56 on other cities. On September 17, 1940, Hitler postponed the invasion, conceding defeat in the Battle of Britain.

At secret locations in both England and Germany, the first aircraft jet engines were being tested. The first flight of an aircraft powered by Frank Whittle's engine took place in May 1941. In Germany, at the then secret Peenemunde site, the ram-jet powered pilotless V-1 was well advanced, and work on the ballistic rocket weapon V-2 was also underway. The first of over 4,000 V-2s was test-fired in 1942; two years later these ballistic missiles were exploding on their targets in England. The British had suspended the TV broadcasts that were started some years before the outbreak of war, but they were using radar arrays to detect German aircraft attacking across the English Channel and Southern North Sea. The world's first programmable digital computer was assembled by Alan Turing in Bletchley Park in England and was already in regular use, decoding encrypted German military communications. And, of course, the nuclear age was being ushered in. The process of nuclear fission was explained in 1939 by the Austrian physicist Lise Meitner and her nephew, the (naturalized) British physicist Otto Frisch. So the movements that would eventually lead to establishment of the Manhattan Project in August 1942 were already underway. I mention all of these to put the rest of the story in proper context. The time is 60 years ago, but many things that are now common features of our everyday life were already in use or under development, such as jet travel, rockets, cruise missiles, telecommunications, computers, and atomic energy. This is not the Dark Ages we are talking about, and yet. . . .

At the end of December 1940, a man named Albert Alexander was admitted to the Radcliffe Infirmary in Oxford, England. Albert was a 43-year-old police officer. I have a mental picture of him that may or may not be accurate: a big man, confident, big-hearted, and a really old-fashioned community cop. He'd been injured about a month before admission to the hospital, though the record does not say whether the injury was suffered in the line of duty. The injury became infected: the dreaded blood-poisoning, or septicemia. Pathogenic bacteria were spreading through the tissues of his face, head, and upper body, growing faster than his immune system could cope with and causing suppurating abscesses all over his face and forehead. His eye became infected and had to be removed on February 3, 1941. Then his lungs became infected. He was close to death and was selected for experimental treatment with partially purified penicillin. On February 12, he was injected with 200 milligrams, followed up with 100 milligram doses at three hourly intervals. The next day his temperature had returned to normal, and he was able to sit up in bed. He continued to improve. But so little penicillin was available that it had to be reextracted from his urine to be reinjected into his veins. Finally, all supplies were exhausted and his condition worsened rapidly. Despite the early promise of cure in what was one of the first cases to be treated with purified penicillin, there was not enough of the wonder drug to save Albert Alexander. He died on March 15, 1941, killed by an injury which had become infected. And the injury which had felled this policeman? Slashed with a knife? Downed by a gun shot? Crippled in one of the acts of war which the civilian population was then increasingly experiencing? No, none of these potentially heroic ends. Albert Alexander had been scratched by a rose thorn.

More than any other scientific advance of modern times, the penicillin story is able to stir interest and imagination because this metabolic product of a common green mold is able to cure a variety of infections that yield slowly or not at all to other treatments. This one fungal product had revolutionized medicine. It cannot cure everything, but penicillin can be used successfully to treat pneumonia, gangrene, and gonorrhea. Diseases such as septicemia and osteomyelitis that were fatal or debilitating, and also widespread, when penicillin first came into use, have been relegated to medical history. In the 1930s (and before), you didn't need a war to be in constant danger of life-threatening infection. Any injury in which the skin was broken might become infected with soil or air-borne bacteria, and these bacteria sometimes grew beyond the level with which the immune system could deal. When this happened, the bacteria consumed the patient and produced toxins in the blood stream that caused more widespread damage. An ordinarily active adult might suffer scratches or minor cuts while gardening, walking, or climbing. Osteomyelitis was an infection of the bone with staphylococcus, which was relatively common in children; how many bloody injuries did your shins suffer when you were a child? Any man who shaved regularly might suffer infection of the inevitable nicks and cuts. This is what killed Lord Caernarvon in 1923, who was a veteran of 19 years in Egypt before he found Tutankhamun's tomb in 1922. And it wasn't "The Mummy's Curse" that finally saw an end to the noble Lord, it was septicemia, caused by an infected shaving cut. Less dramatic infections caused unsightly boils and abscesses all over the face as bacteria established in shaving cuts invaded otherwise healthy hair follicles. Not deadly, but painful, debilitating, and above all, common. And women were even more at risk. Birth is a very messy procedure even today, but before the ready availability of antibiotics, an astonishing proportion of new mothers suffered, and countless died, from puerperal fever resulting from internal infection. And an equally astonishing number of newborns were infected during birth with bacteria that were only mildly pathogenic but nevertheless caused blindness, deafness, and other lifelong disabilities, even if the child survived. Bacteria are very successful organisms. They are widespread and adaptable. Life for humans was not easy. Penicillin contributed in a major way to a revolutionary change in medical treatment that in turn changed the human lifestyle to such an extent that diseases that were common causes of death and disability are now rarely encountered. This dramatic change to the everyday experience of everyone on the planet is to me the most remarkable aspect of the discovery and introduction of penicillin. There are several other remarkable aspects of the penicillin story, supply being one of them. For Albert Alexander, one of the first patients to be treated, there was little more than 1 gram available in the whole world, and it was not enough to save his life. Today we produce enough of the totally purified material to give every human on the planet a dose of 5 grams!

Development of penicillin production on an industrial scale was a triumph in both scientific and technological terms. Penicillin was discovered, apparently accidentally in 1928, at St. Mary's Hospital in London by Alexander Fleming. There are a number of different accounts of this discovery, but the usual story is that culture dishes of the pathogenic bacterium, *Staphylococcus aureus,* which Fleming had left on his work bench, became contaminated with a fungus spore. When Fleming returned from his brief holiday, a fungus colony had formed, and growth of the bacterium was inhibited in a zone around the fungus. Fleming identified the contaminating fungus as a species of *Penicillium* and named the unknown inhibitory substance penicillin. Fleming studied the material to some extent during the 1930s but was unable to purify or stabilize it. He clearly recognized its potential, suggesting that penicillin might have clinical value if it could be produced on a large scale.

Penicillin was later purified by Howard Florey, Ernst Chain, Norman Heatley, and other members of a team at Oxford University, which is how the antibiotic became available for experimental use at the Radcliffe Infirmary in Oxford in 1940. The Oxford team developed a purification method using the solvent ether, which gave good yields of antibiotic from a meat broth medium on which Fleming's fungus had been grown. The result was a brown powder that was a remarkably potent antibacterial agent, even though still not fully purified. The essence of the production process was a surface fermentation method in which the fungus was grown like a crust on the liquid medium. To produce enough penicillin-rich cultivation liquid, the Oxford group used all types of readily available bottles and dishes, but what proved to be the best of these makeshift containers were hospital bedpans! Later purpose-made ceramic or glass vessels modeled on this utensil were renamed "penicillin flasks".

Despite heroic efforts to scale-up production, the British fermentation industry simply lacked the knowledge and expertise to produce penicillin effectively. American academic and industrial scientists were well ahead of their counterparts in the rest of the world. They had experience of growing fungi in deep fermentation submerged culture and were experts in the selection and development of high-yielding strains. Indeed, the crucial American contribution to industrializing penicillin production grew out of the experience of American scientists during the 1930s, like Selman

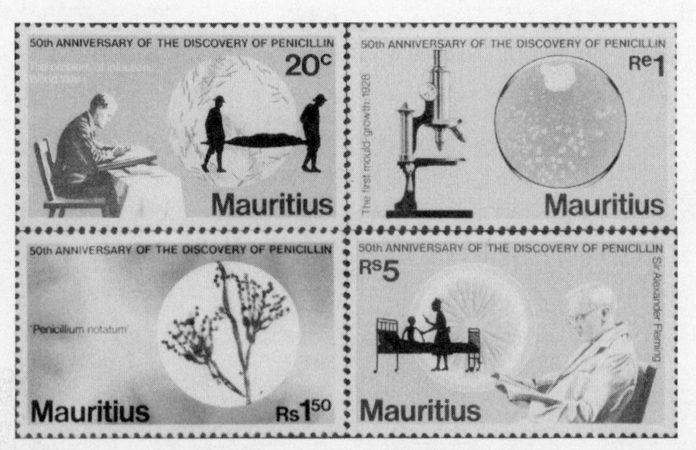

A set of stamps issued by Mauritius to commemorate the 150th anniversary of the discovery of penicillin gives a very succinct outline of the story.

Waksman and Harold Raistrick, in the usefulness of fungi for industrial production of chemicals such as fumaric acid and especially citric acid. Transfer of penicillin development to the United States enabled the marvel of wartime antibiotic production to be achieved. By the end of World War II, penicillin cost less to produce than the packaging in which it was distributed. It made an enormous contribution to the war effort, too. During World War I, 15 percent of battle casualties died of infected wounds. When penicillin became widely available in the latter half of World War II, recovery rates from nonmortal wounds were routinely 94 to 100 percent; death from infection of the wound was almost zero.

Of course, this medical and military advance was not shared with the enemy. The armies of the Axis Powers continued to suffer the additional slaughter caused by infection of wounds. Even after the war, although transfer of American know-how in fungal fermentation enabled Britain to become self-sufficient in penicillin production by 1947, much of the rest of the world remained dependent on U.S. production for a long time. Remember that the plot of Graham Greene's classic book and film *The Third Man* centered around Harry Lime, a black market supplier of penicillin in post-war Vienna. Penicillin was so important (and valuable) that it was worth risking death to smuggle.

From every angle the development of penicillin is a remarkable story. The drama is heightened by its timing—right at the start of a worldwide conflict—and by the fact that this was the first time that a medically useful fungal product was subjected to industrial-level production. Yeasts have been used since antiquity to leaven bread and to produce alcoholic drinks like beers and wines, and several traditional foods have been flavored with fungi.

In ancient times fungi were used a great deal for their supposed curative properties, but, in the Western world at least, such use declined until interest revived in the potential applications of fungi and fungus metabolic products in medicine following the success of penicillin. Declining use of fungi and plant materials can be traced to the application of the reductionist approach of the Western scientific method, which made it difficult to sustain the varied claims made for many natural products in traditional medicines. One of the writers on this topic claims that ancient writings from all civilizations refer to the use of molds to treat infected wounds. There seems to be good evidence that our ancestors maintained cultures of therapeutic molds or knew how to grow them selectively so they could be used to cure surface infections. The practice was revived in the early 1940s soon after penicillin was identified but before the purified product became widely available.

Much of the traditional usage seems to have escaped formal written record although Herbalists and other medical writers of the middle ages included fungi (mainly mushrooms and toadstools) for their abilities to cure various complaints. Like most other aspects of Western science of the time, these writings owe a lot to those of the Greeks more than 1,000 years before. For example, Dioscorides, writing around 200 A.D., claimed that fungi could cure a great many ailments and amounted to an almost universal remedy. Extravagant claims like this are readily found also in Chinese and Japanese traditional medicine. For example Lingzhi (fruit bodies of the fungus *Ganoderma*) is described as "the rarest and most precious Chinese medical herb," that legend claims even to have "the miraculous power of raising the dead to life." Less extravagant claims are for Lingzhi "preventing and mitigating a variety of clinical conditions: chronic bronchitis, asthma, neurasthenia, insomnia, amnesia, hypertension and hypotension, coronary heart disease, arrhythmia, stroke, hyperlipidemia, thombosis, female endocrine disorder, female physiological disease, menstrual disorder, chronic hepatitis, gastric diseases and duodenal ulcer, allergic and chronic rhinitis, dysuria, arthritis, rheumatism, allergic dermatosis, cancer." Now that quotation does not come from some ancient medical text, but from a leaflet I picked up in a departmental store in Hong Kong in April of 1999! So the material is being sold now to sophisticated and highly educated people, to the engineer who designed and built your lap-top computer or the pilot of your next trans-Pacific flight, for example. And it's being sold on the basis of a written medical tradition that goes back more than 5000 years. That a similarly ancestral, but sadly unwritten, tradition occurred in Europe is indicated by material carried by the Alpine traveler who has become known as The Iceman.

Around 3200 B.C., a Neolithic traveler set out across the Alps. He didn't make it. Somehow he was caught in the ice and snow, entombed and preserved in the glacier. Eventually, as a result of the glacier's slow descent of the mountains, his corpse was exposed at the edge of the ice sheet in 1991 close to the Austrian–Italian border. A well-preserved 5,000-year-old corpse with all of its clothes and equipment is a remarkable find by any measure. But possibly most remarkable is that there were three separate fungal products among the Iceman's equipment. One of these is easy to

Ganoderma **seems to be naturally variable, growing as a bracket on upright logs but with a stem on horizontal ones (top photographs). Color is also variable (bottom photograph), with some colors meriting higher prices than others.**

account for. It was a mass of fibrous material in a leather pouch together with flints and a bone tool like an awl. This fungus has been identified as one with a long history of use as a tinder, so clearly it was part of the Iceman's fire-making kit. The other two are more problematical. Both are pieces of a bracket fungus (actually called *Piptoporus betulinus*) and both are threaded onto leather thongs. One piece is essentially conical, about 5 centimeters in its longest dimension, and is on a simple leather thong. The other is spheroidal, with a 5-centimeter diameter, and is on a thong that has a lobed tassel at one end. These objects were clearly carefully made and must have been important to the owner to be included as part of the kit he chose to take with him in his trek across the mountains. *Piptoporus* is known to produce (and accumulate in its fruit bodies) antiseptics and pharmacologically active substances that are claimed to reduce fatigue and soothe the mind. I can imagine that with due ceremony and additional magic, these objects may well have been seen as essential to the traveler in the mountains. The conical one could well be a sort of styptic pencil to be applied to scratches and grazes, and perhaps the flattened, spheroidal one was chewed or sucked on when the going got tough.

So our distant European ancestors held fungal products in such high esteem that they were necessary accessories for hazardous journeys. What of today, and what of the future? Today, alcohol and citric acid are the world's most important fungal metabolites in terms of production volume, although penicillin can still lay claim to be the most important. Since the introduction of penicillin, several millions of chemicals and metabolites have been screened for antimicrobial and other pharmaceutical activities. The lesson has been well learned, and chemical screening is a major activity of the pharmaceutical and agrochemical industries around the world. Most antibiotics that we use today actually originate from bacteria, particularly the streptomycetes. Indeed streptomycin was the second significant antibiotic to be found and was largely responsible for the demise of tuberculosis as a major disease (resistance to streptomycin being a cause of the recent resurgence of tuberculosis). Antibiotics obtained from fungi that are presently of clinical use as antibacterial agents include the still-important penicillin, cephalosporin, and fusidic acid (both of the latter are useful against penicillin-resistant bacteria), and the antifungal griseofulvin (used to control fungal infections of the skin, nails, and hair). Obviously, antibiotics are the products that come to mind first when thinking of medically useful fungal products, and they have dominated this chapter so far. But fungi have much more potential, and some products first discovered as antibiotics have subsequently been shown to possess some other potent activity in tests on mammals. For example, the increasing importance of organ transplantation in medicine focuses attention on compounds capable of suppressing the recipient's immune response so as to avoid organ rejection. The fungal product called cyclosporin is now widely used as an immunosuppressant and greatly improves the success rate of transplant operations. Another fungal product, gliotoxin, seems to regulate the immune system and may also be useful for postoperative management of transplant patients. A final example of a natural compound obtained from fungi that has great medical value is a hydroxy acid called *mevinolin.* This is produced by the fungus *Aspergillus terreus,* and it acts as a cholesterol-lowering agent by interfering with enzymes that make cholesterol in mammals. By the mid-1990s, three compounds derived from mevinolin had worldwide sales that put them individually in the top-ten-selling pharmaceuticals by value of sales. These were Pravastatin (sales in 1995 valued at two billion U.S. dollars), Simvastatin (sales of 1.6 billion U.S. dollars) and Lovastatin (sales of 1.4 billion U.S. dollars). All of these "statins" are used to reduce cholesterol levels in the body because high cholesterol levels are considered to be a risk factor in heart disease.

Fungi also produce compounds like the ergot alkaloids, steroid derivatives, antitumor agents, and immunoregulators. I described the terrible effects of the ergot toxins in chapter 1; ergot-contaminated grain caused St. Anthony's Fire; with this condition, the sufferers experienced gangrene, cramps, convulsions, and hallucinations. The toxicological effects are caused by the numerous alkaloids that ergot contains, but in low and controlled concentrations, these are valuable drugs causing dilation of veins and a decrease in blood pressure as well as contraction of smooth muscles. They are drugs of folklore too, for the ancient remedy was to give ergot in childbirth to hasten contraction of the uterus. The ergot alkaloids are now produced by fermentation similar to penicillin fermentation. Although the alkaloids can be synthesized, strain improvement by mutation and selection of high-yielding strains has been so successful that fermentation remains the most cost-effective means of production.

Fungi don't have to produce a compound to be useful because they can bring about a variety of chemical transformations of compounds, especially steroids, in a reliable and reproducible way. Most of the steroids in clinical use today are modified during manufacture in this way. The fungus is grown in the minimum amount of nutrients (this limits complications during subsequent extraction). Then the steroid to be transformed is added, and about 20 to 40 hours later the new steroid can be extracted. The attraction is that this relatively simple procedure may avoid anything up to 20 or 30 steps of pure chemistry. Success with steroid transformation led to similar approaches being applied to other pharmacologically active compounds, to reduce toxicity while maintaining activity, or make chemical alterations to enhance activity. Using fungi in this way is a convenient and economic means of making specific compounds that would be otherwise very difficult, impossible, or just too expensive to produce by direct chemical synthesis.

Cancer would be high, if not top, on most lists of afflictions in need of a reliable cure, and consequently, a key feature of most screening programs is the search for effective and safe antitumor and antiviral agents. Several hundred antitumor agents have been isolated from microbes, most of them from bacteria. Nevertheless, many fungal products have been found to inhibit the growth of cancers in animal tests. Specificity and safety are the issues which limit the medical usefulness of most of these compounds; they may have adverse effects on the host organism as well as the tumor. The most successful antitumor agent is a polysaccharide isolated from the shiitake mushroom. It is called lentinan and seems to work by modifying the activity of the patient's own immune system, making it more active against the cancer cells. Lentinan has been used successfully to treat stomach cancers and is now a multimillion dollar industry. Since cancer control was one of the benefits claimed for shiitake in the ancient

Chinese and Japanese medical writings, the successful development of lentinan prompts the hope that the (many) other fungi used in traditional Oriental folk medicine may also yield products useful in today's clinical practice. The search is on. Over a period of several thousand years, an enormous number of fungal products have been recognized as having medicinal value in China, Korea, and Japan. Oriental science and Western science are very different. A Western scientist seeks to isolate active principles from one another and to understand the parts individually and separately. Oriental science is holistic, almost anecdotal in Western terms, because it seeks to understand the entire complex event. We might hope, therefore, that at least some of the oriental remedies might yield to modern Western analytical techniques so that the active principles of centuries-old medicines can be identified, purified, and turned into "Western-style" drugs.

Of course, if you have faith in oriental wisdom you can use the traditional remedies right now. The term *mushroom nutriceutical* is used for a new class of compounds extractable from various mushrooms. Mushroom nutriceuticals can have both nutritional and medicinal properties. Nutriceuticals extracted from medicinal mushrooms have extraordinarily low toxicity, even at high doses, which is an advantage over most pharmaceuticals. Long used as tonics in traditional oriental medicine, they are now believed to profoundly improve the quality of human health. There are claims that consumption of mushrooms as a food or the use of mushroom extracts as dietary supplements can enhance the immune response of the human body. The expectation is that this would increase resistance to disease and even have the potential to cause regression of an established disease. Several fungi have been found to produce metabolites that inhibit the multiplication of viruses in culture and in animal tests. Activity of various of these compounds against major disease-causing viruses, like poliovirus, coxsackievirus, vaccinia, and various influenza viruses has been demonstrated in the laboratory, but the tricky transfer from laboratory curiosity to clinical tool has not yet been achieved. Only time and further research will tell if we can exploit the promise of the Chinese fungal remedies.

6

Turning the Tables

Using Fungi to Control Other Pests

Just about everything we do is affected by pests of one form or another, and over the years we have invented an armory of chemical pesticides that have allowed for enormous improvements in agricultural and horticultural yields. However, most of these chemicals are powerful and indiscriminate poisons, and worries over the adverse environmental impact of heavy usage of chemicals like these are increasing. As we have seen so far in this book, fungi are very effective pests of other creatures in their own right, so it's not surprising that attention is turning to the use of fungi as control agents by harnessing their natural antagonisms to pests of our crops, and there is potential for us to use fungi in controlling other fungi that cause diseases of crops, as well as insect pests, nematode worms, and even weeds.

Over 2,000 years ago, the Chinese were writing about the fungal diseases of silkworms and cicadas. So the importance of fungi that infect insects (they're called entomogenous fungi) in natural populations of insects has been recognized for a long time. It's interesting that early support for the germ theory of disease came from a fungus disease of silkworms. The disease was called muscadine disease, and it killed silkworms in Europe so effectively that the future of the silk industry there was in peril. A lawyer-turned-farmer named Agostino Bassi found that the silkworm corpses produced masses of fungal spores. The fungus, now called *Beauveria bassiana* in Bassi's honor, is one of today's prime candidates as an alternative to chemical pesticides. Bassi is worthy of honor. The people who are remembered as establishing the germ theory of disease are Pasteur, Lister, and Koch. Great men indeed, but they did their work in the 1870s, and Bassi had already settled the issue in 1835–1836 (10 years before Berkeley's work on the late blight of potato, discussed in chapter 2). Unfortunately, choosing to work on insects and fungi is a sure way of being forgotten, then as now, and Bassi never got the credit he deserved, although they did name the fungus after him! The idea that these fungi might be useful as biological control agents was first explored by two pioneering scientists in the Ukraine, Metchnikoff and Krassilstschik, in the 1880s. These two visionaries mass-produced the spores of an insect-attacking fungus (called *Metarhizium*) and tested it against insect pests of wheat and sugarbeet. Throughout the twentieth century, insect-infecting fungi have been assessed as possible control agents for a variety of insect pests. There's a repetitive pattern. Field experiments are carried out, and products are developed and may even be successfully marketed for several years. But then the fungal products are replaced by more effective chemical insecticides. The sequence of development, commercialization, and later withdrawal has been repeated with a number of other fungal products. Currently, there are no fungal products that are widely used for insect control, although in particular parts of the world (including Russia, China, Brazil, and the United Kingdom), and, more importantly, for particular insect pests in very particular circumstances, there are fungal products in use.

There has been a renewed interest in these insect-attacking fungi because of increasing insecticide resistance and environmental concerns over pesticide use. New strains of fungi have been isolated from a wide range of hosts, and these have even further emphasized the potential that exists for the use of fungi in insect control. But the continuously repeated cycle of development of a promising fungal product followed by its eventual failure has emphasized the importance of ecological features in determining their success as biological control agents. The problems encountered relate to the biology and lifestyle demands of the fungi, which we try to use as control agents. Over one hundred different sorts of fungi (representing all the main fungal subdivisions) have been shown to parasitize living insects. They are more common in tropical areas where temperature and humidity favor their growth. But they occur in most ecosystems, ranging from natural water habitats to the high technology horticultural production systems we create around the world. The host range of individual fungi is variable; some attack only one insect and others, such as *Metarhizium*, have a broad host range.

The key controlling feature that determines the usefulness of these natural control agents is their dependence on humidity for germination and growth. For them to attack their insect prey. their spores must germinate on the insect and the processes of spore germination and growth of the spores on the surface of the insect are highly dependent on both available moisture and temperature. Even 90 percent relative humidity might result in only half the spores surviving. In the real world this means that the microclimate in the tiny area that the host insect prefers might determine the success or otherwise of the control agent. For example, in attempts to control aphids in glasshouses, there was a major difference in efficiency against aphids that fed on the undersides of leaves (very effective control) where the humidity is high, compared to the more exposed stem (virtually no control). In similar experiments, the spread of infection through aphid populations was maximal when free water was present but completely prevented at 93 percent relative humidity. Still rather damp to you and me, but dry as a desert to the fungal spores. Temperature also affects germination and growth. Both are markedly reduced below 15°C and above 35°C , so the window of effectiveness is fairly narrow.

Once the fungal spore has germinated and grown through the insect's skin, it produces cells that circulate within the insect and proliferate by budding. Insect death usually occurs 3 to 14 days after the fungal spore alighted on it. Death results from a combination of mechanical damage resulting from tissue invasion by the fungus, loss of nutrients to the fungus, and reaction to toxins generated by the fungus. After the death of the insect, the fungus grows on its corpse, eventually producing a new crop of spores. So the initial application of fungal spores results in the production of even more spores after the first successful infections. It is this amplification of the control agent that makes the biological control mechanism so attractive. A chemical agent will be diluted, washed away in the rain, or used up. But as long as there are hosts to infect, the biological control agent will grow stronger and stronger, producing an epidemic that can destroy the target insect completely. Unfortunately, this promise is rarely realized, though there are sufficient successes on the record for us to keep trying.

Rice pests are potentially good targets for these insect-attacking fungi; the high humidity in rice paddies and warm temperatures in rice-growing areas combine to produce close-to-ideal conditions for epidemic infection of the insects. Attempts to control rice pests in the field by applications of fungal spores have indeed been relatively successful. Pests of glasshouse crops can also be controlled well by biocontrol agents. This is probably the most ideal circumstance because the environment within the glasshouse can be controlled to optimize the infestation. Even here, though, the environmental demands of the parasites can cause difficulties. For example, control of whitefly on cucumber crops is effective only after the crop is four-weeks old. This is because the smaller leaves of younger crops greatly reduce the humidity around the plants, and the fungal spores cannot germinate. The age of the insect pest can influence results, too. Ninety-four percent kill was obtained when eggs of the white fly were treated (because the first larvae are then infected as soon as they emerge from the egg), but only 28 percent of older larvae were killed when the treatment was delayed. The fungal spore preparation has to be applied several times to maximize, and the spray must cover the entire plant to get good spread through the whitefly populations because the larvae themselves are not sufficiently mobile to spread the fungal disease. If all these requirements can be met, the development of fungal infections in the whitefly when humidity is high is very rapid, and it is possible that a single night of high humidity in a glasshouse would be enough to destroy a whitefly infestation. There are no chemicals and no chemical residues, so biocontrol of whitefly, although difficult and demanding, is very promising.

Another successful application of fungi is in the biocontrol of the lucerne aphid in Australia. This pest was introduced to Australia with none of its natural enemies, and it rapidly became a major problem. A search for natural lucerne aphid disease agents found several likely pathogens in Israel. One of these was introduced into Australia; it spread rapidly and now helps to maintain the pest population at a low level.

Beauveria bassiana, that belated monument to Agostino Bassi, has been tried as a microbial control agent of several insect pests in different parts of the world, although most development work on this organism was done in the then Soviet Union, where it was mainly used for control of Colorado potato beetle. Field trials showed that it was more active against weakened insects, and the standard treatment developed was to use the fungus in conjunction with a quarter of the usual rate of insecticide application. The reduced dose of insecticide weakens the insects to such an extent that the fungus can easily infect and kill virtually all of them. In the 1970s through the 1990s, preparations of *Beauveria bassiana* spores were used on several crops over an area of at least half a million hectares in China. The fungus was used to help control pests like the corn borer, pine caterpillars, and leafhoppers. Biocontrol agents of this sort are particularly useful in peasant farming and commune-based agricultural systems because the community can produce its own supplies of fungal spores. In this case, spores were produced by growing the fungus on boiled rice. This puts production of the control agent into the hands of the community that needs it in a way that simply cannot be done with chemical insecticides, which usually have to be bought with hard-earned cash from multinational chemical companies. This empowerment of the community is such an attractive proposition that it may outweigh any lesser effectiveness of the biocontrol agent when compared with the chemical treatment. There's a similarly successful example from Brazil. The fungus known as *Metarhizium* is used commercially in Brazil to control spittlebug of cane sugar plants. The fungal spores are produced by growing the fungus on sterilized rice. Once a good crop of spores have appeared, the rice grains are dried, grains and spores are ground up, and the powder is sold under a variety of trade names. Again, this process is very simple and is well suited to production on a local basis by grower cooperatives. The yield of spores is very good, about 1 million million spores per kilogram of rice, which is enough to treat a whole hectare of cane sugar. Spittlebug control is one of the few major success stories in biocontrol. Conversely, trials of the organism *Beauveria bassiana* in the United States have been less successful. During a three-year test, control of the potato beetle was highly variable. In only 8 out of 24 trials was potato yield from fungus-treated plots significantly greater than yield from control (untreated) fields, but only 2 of these trials of these gave yields on a par with insecticide treatment. It seems likely that the more industrial farming practices result in high pest numbers at the time of fungal application and that this reduces the effectiveness of the fungus.

As these stories show, there have been several major successes in particular places, under particular circumstances, and with particular pests. These successes are a spur to continued research, but there has been no really dramatic breakthrough, nothing that has wide effectiveness and can be used under a range of conditions. What is most urgently needed is a fungal formulation that is not only safe to use (the common characteristic of these agents), but also cheap to produce, easy to apply, not too dependent on specific environmental conditions, and that can consistently control the target pest.

Unfortunately, we are not very close to being able to achieve that. There are no commercially available mycoinsecticides in the United

States or Europe that satisfy this need. The history of the topic is littered with products launched with high hopes and then withdrawn because they failed to provide reliable pest control. Astonishingly, despite this disappointing experience, commercial interest continues, and it is remarkably widely accepted that fungi have the potential to control insects, aphids, and other pests. It seems to be a topic in which blind optimism wins out over bitter experience. I suppose that to a great extent, this blind optimism is driven by the possible advantages of mycoinsecticides compared to conventional chemical insecticides. Fungi infect most sorts of insects, aphids, and mites, but individual strains of fungus can be very specific and may only infect one type of host. So, you can imagine fungi being used to control important pests without affecting innocuous insects in the same environment; that's part of their inherent safety. Of course, insect-attacking fungi cannot infect any other type of animal (or plant, for that matter), so unlike some of the more toxic chemical treatments, the environmental impact of a mycoinsecticide is strictly limited to the host species it was developed to control. Another advantage over chemicals is that there is no evidence so far of fungal resistance occurring in insect populations. Now, this might be due to the fact that they've been used to only a limited extent so far, but it is a sharp contrast with chemicals that seem to select out resistance strains of the pest very quickly. The holy grail, therefore, is that widespread mycoinsecticide use could provide great benefits to us and our environment. Like the real holy grail, the problem is to find it.

The generally disappointing experiences of the past hundred years or so, and, to be fair, the few successes, do give some indications as to what directions we need to take in the future to realize the high hopes that so many people have for mycoinsecticides. The first point is the crucial importance of selecting the right strain of fungus. An enormous number of fungi exist and just about all features of the pathogen–host relationship can vary greatly between strains of any one species. By searching through a large collection of different strains, one with the desired combination of characters will eventually be found. Molecular biology might help.

Genetically-modified pathogens of pest insects might be designed that bring together toxins, virulence, or other pathogen functions from organisms that cannot be mated together conventionally. Progress is being made in understanding pathogenicity mechanisms, particularly things like how the fungus first penetrates the outer defenses of the host, so it might become possible to increase the speed of insect kill by genetic manipulation of the fungus. This is important because most mycoinsecticides can only be applied after the crop has become infested by the pest. That being the case, it is essential that the pest is killed rapidly, otherwise you'll end up with the unsatisfactory situation of killing the pest after it's done its damage.

Once you've selected, engineered, and genetically modified a really mean insect pathogen, you need to be able to produce it on a large scale if it is to become a commercial proposition. The most attractive production method is liquid fermentation using stirred tank reactors. The engineering of this technology is well understood because it is used for other useful products like citric acid and antibiotics. However, as I've explained previously, two of the most successful examples used a method using cereal grains as a semisolid medium to produce the fungal spores. This is a traditional way of producing several human food products, so it is a type of production that offers possibilities. However, each new product requires new research, so it can be a slow development program. Another commercial requirement is that the product should be easy to store. One expert has suggested that a storage life of more than 18 months is necessary for a commercial product. Few farms around the world can afford refrigerated and/or humidity-controlled storage, so we're talking about storage in a barn or shed with no special precautions. These aspects can be controlled by the nature of the formulation with which the fungus is mixed to make the commercial product. The formulation can also assist initial infection if it includes materials that help maintain favorable humidities while the spore germinates on the host. There is a limit to this, of course, and then the product is at the mercy of the real world, depending on atmospheric factors and the density of the host population to achieve the desired catastrophic infection of the target organism, prolonged pest control, reduced risk of resistance, and a high degree of safety to nontarget organisms with all the associated environmental benefits. It's a long job.

It's not only insects that are pests. Weeds pose a large problem in both agricultural and natural ecosystems, not just in gardens and amenity sites in urban areas. Most people would agree that an insect that eats its way through the leaves of a plant is something of a pest, but weeds are not that easy to spot. What is a weed to one person is a wildflower to another. A plant might be designated as a weed because it has a detrimental affect on agriculture or amenity plantings, but the same plant may actually benefit beekeepers, wildflower and fruit gatherers, herbalists, and others. Not surprisingly, attempts to control the so-called weed may generate heated arguments. There's an interesting story about a plant called *Chromolaena odorata* from Brazil, which became what is known as a pantropical weed, and caused a biologically devastating invasion of Ghana and the Côte d'Ivoire in 1993. In Brazil, where the plant was native, it was not considered a weed, and though classed as common in Brazil, it never formed the dense, continuous stands that characterized its growth pattern in West Africa (and Asia). The plant caused significant agricultural and environmental damage over a large area of West Africa and Asia and was suggested as a good candidate for a biocontrol program. This led to an outcry, however. Some people suggested using chemical herbicides, despite the enormous area that would have to be sprayed. But others argued that the weed should not be controlled at all because the plant had medicinal properties and because it formed a rapid plant cover in the slash and burn agriculture of the region. The counterarguments were so strong that funding was withdrawn from the proposed biocontrol project.

Agricultural losses due to weeds are difficult to estimate, but there is broad agreement that at least 10 percent of the world's agricultural production is lost each year. For the United States, this means that annual crop losses due to weeds range up to 30 billion U.S. dollars. Estimates like this do not include the negative effects of environmental weeds in natural ecosystems. Another measure of the importance of weeds is to estimate how much people are willing to spend to get rid of them by using herbicides. Herbicide sales account for about 70 percent of all pesticides sold (the proportion was only 20 percent in the early 1950s) and have a retail value of around 20 billion U.S. dollars. Invasions by weeds that are foreign to a region have become a particular problem as international travel and commerce have increased. Such introductions (usually unwitting and unintentional) can invade native vegetation and outcompete the resident plants so disrupting the balance of the whole ecosystem.

The recognition of fungi as important natural enemies of weeds that might be safely exploited to our benefit is not new, but it wasn't until the 1970s that they began to be used in serious attempts at weed biological control. The approaches used for control of insect pests are just as valid for biocontrol of weeds with fungi. Where the target is a plant that has become a pest outside its natural range by being introduced into some so-called exotic area, then the best strategy is to introduce a fungal natural enemy of a target weed from its native range into the exotic area. An alternative strategy is to use a fungal pathogen from the exotic area but to mass produce it and apply it at such a high concentration that the weed population is inundated with fungal disease. Just as with attempts to control insect pests, it is a long job and is dependent for success on enormous amounts of scientific research. It is also essential to make proper economic evaluations of the project. No matter how attractive the idea might be from scientific and conservation points of view, someone will have to foot the bill and economic feasibility, and cost/benefit analyses must be done.

At first glance the use of fungi in weed biocontrol seems like enlightened management, which is likely to solve important problems and contribute toward sustainable agricultural development. But the truth is not quite so clear cut. Yes, you can find numerous examples of weed infestations that have caused major problems like disrupting fishing activities, harboring disease vectors, and major crop losses. In some instances the effects have overtaken whole cultures and led to agriculture being abandoned or major population movements to escape the problem. No one can deny that weed control (with fungi or with chemicals) in such dramatic cases is desirable in social and cultural terms. But these are the dramatic extremes which are, thankfully, rare. In most places, most weeds are just a few plants growing out of place in a field or other cultivated plot. Careful husbandry should be able to cope with them. In fact, weeding is still one of the most common activities of a large part of the human population. So think of it this way: in a sense, weeds could be seen as being beneficial because they generate honest employment. You might counter that by pointing out that physically removing weeds is such a demeaning human activity that it is a waste of human resources. Freeing the human population from this burden then becomes a noble goal for scientists and technologists alike. The attractions of the "honest toil" argument are deceptive, indeed illusory, and are directly analogous with the development of industrial machines that change the nature of the industrial shop floor. Supplementation, and eventual replacement, of human toil is the desirable aim, and herbicides have the capacity to do that. When mycoherbicides are common in the marketplace, they will expand the options available to major users of herbicides.

Control of weeds with mycoherbicides is an ecologically clean method. It uses a natural agent and leaves no chemical residues. There are concerns that fungal plant diseases that are used as mycoherbicides might also pose a threat to crops or native plant species. Knowledge of host specificity, disease severity, and possible tolerance to the disease is imperative. Paradoxically, a fungus that is capable of attacking a native or even a crop plant may have potential as a biocontrol agent under particular circumstances. It's then a matter of weighing the cost of causing disease in native plants against the benefit of using that same disease to control the weed. If the weed population is less tolerant of the disease, then the balance shifts in favor of using the fungus to control the weed. This type of consideration also applies to chemicals. Most pest control chemicals are toxic to a wide range of organisms; the useful ones are more toxic to the pest than to the crop. It's again a matter of scientific research. The risks of plants other than the target weed being affected by an introduced mycoherbicide fungus can only be judged if appropriate scientific knowledge about the mechanisms of fungal host-specificity is available.

The potential economic value of mycoherbicide is becoming easier to judge as more trials are completed. Introduction of a rust fungus into Australia to a weed that originated from the Mediterranean region was certainly biologically effective. It resulted in better than 99 percent reduction in infestations by the weed. But the estimated annual saving due to increased crop yields and reduced chemical herbicide use was 16 million Australian dollars. Since the cost of the whole project was only 3 million Australian dollars, the return on this initial scientific investment has been enormous. Unfortunately, benefits on this scale are not often seen. The mycoherbicide business has a history, similar to the mycoinsecticide business, of products being researched, brought to the market, and sold apparently successfully for several years, but then being withdrawn. Often the market is too limited to repay the high initial development costs and then support the ongoing costs of further development to cope with competition. Several products that were successfully sold in the 1980s were withdrawn in the 1990s because of the costs of registration procedures imposed by environmental protection agencies.

This last point brings home the fact that a major obstacle to the introduction of biological control is (unjustified) fear of the spread of disease. This pathophobia has led to overzealous rules and seemingly endless tests and trials. Oddly enough, this bias against introduction of fungal plant diseases as biocontrol agents is not also applied to introductions of insect pests, and most countries pay little attention to precautions against new weed introductions. About half of the weeds in the United States and 13 of the top 15 weeds in the United States are introduced species. Similarly, 78 of the 107 most noxious weeds in Canada were introduced to the country. The best way to control such aliens is to introduce diseases from their home territories. But the regulatory authorities make that process unreasonably difficult (and unrealistically costly). This, despite the fact that the overall rate of effectiveness that has been achieved with the fungal diseases introduced in the past, is 67 percent.

Plant parasitic nematode worms are also candidates as targets for using fungi as biological control agents. These nematode worms spend part of their life cycle in soil or on the root surface where they are exposed to nematode-attacking (called nematophagous) fungi that could provide useful alternatives to chemical nematicides. Current evidence is that trying to enhance the activities of resident nematophagous fungi in soils requires too much material to be generally acceptable. On the other hand, adding biological-control fungi to soils where they are scarce or absent could lead to commercial products. But many problems remain to be solved before predictable control can be achieved at practical rates of application. If nothing else, greater knowledge of the pest and its interactions with soil, host, and fungal disease agent might lead to identification of more acceptable chemical nematicides.

One thorny problem that all biocontrol projects face is that of what to do about patenting the biocontrol fungus. Commercial companies would expect to be able to get some patent protection for any agent in which they have invested. A chemical is relatively easy: the chemical itself, the way it is produced, and the entire family of its derivatives can be patented. The material can then be put on the market, with the confidence that even if a competitor buys a supply of the material, they cannot make more of it very easily. When the agent is a live organism, however, any competitor with a half-way competent biologist would be able to grow more of that organism from even the smallest sample of the commercial material. If the biocontrol agent is a genetically modified organism, its identity is not likely to be difficult to establish, and the company that first created it will, therefore, be able to identify and claim what rightly belongs to them. But suppose the biocontrol agent is a fungus that was isolated from nature. Say, for instance, staff of a company or institution collected fungi from a particular site and found one of them to be especially virulent in controlling a specific pest. What's to stop a competitor going out to the same place and finding a related fungus that's equally virulent? What's to stop an unscrupulous competitor growing up that fungus and simply claiming that it's something they've newly found? There's no difficulty in writing patents, registrations, or other legal devices that assign ownership. The real difficulty is in having a catalog of identifying features of the commercialized organism that is sufficiently good to ensure that one could recognize it in any competitive product. Genetic markers help a lot with organisms, like fungi, that have few distinguishing features of their own. That's why there's so much interest in genetic fingerprinting. These techniques really would allow one to identify that a competitor had stolen an organism. Of course, you can never stop competitors going out to find a biocontrol organism of their own. The thing to do in that case is to license your patented product for your competitor to sell at a price less than the cost of developing a separate competing organism. Badge-engineering for fungi!

7

Let's Party!

Well, we've done a good deal of work up to this point so why don't we have a break? I'm enjoying your company, let's go for a meal, my treat. I know a nice little place not far from here: *Valentino's* they call it. You'll like the place, it's small and quiet but serves an amazingly wide range of dishes.

Now, The first question is: what are you going to have to drink? Unless you choose a still fruit juice, the chances are that some fungus or other will have contributed to its production. All those Fizzy drinks contain citric acid, and a fungal fermentation process makes most of it, with a global production of around 300 thousand metric tons each year. But the really important fermentation, of course, is the one that makes alcoholic drinks. It's a simple process. Start with something that's rich in sugar, like fruit juices, honey, or cereal grains or roots. Then add yeast (or rely on the yeasts already in the starting material) and allow the mixture to ferment. You'll end up with a liquid that contains up to 16 percent alcohol! Drink it fresh or, if you have the patience, let it age for a while or, if you really want to pickle the nerve endings, distill it into a spirit.

It's such a simple process that all, even the most primitive, societies have one or more fermentation processes that they include in their rituals. There are some beautiful ancient Egyptian murals and tomb ornaments depicting bread and wine making. I suppose it says something about the human condition that alcohol (and caffeine, for that matter) has been incorporated into the way of life of every civilization. From the biological point of view it's remarkable that the organism responsible for fermentation is invariably the yeast called *Saccharomyces cerevisiae* (not surprisingly known as Brewers' yeast) or some closely related variant. The yeasts used for making ales tend to form a froth and grow on the top of the mix, and *S. cerevisiae* itself is such a top-fermenting yeast. Lager yeasts do their fermenting at the bottom of the tank, and they belong to the related *S. carlsbergensis.* Wine yeasts are an elliptically shaped variant of *S. cerevisiae* called *S. ellipsoideus.* Cider yeast is called *S. uvarum,* and saki yeast is called *S. sake.* There are so many yeasts around, and it's remarkable that the alcohol-producing ones belong to such a closely related little family. And they support massive industries; our annual global consumption of alcohol is currently 30 billion liters. That's a lot of yeast!

We don't drink it all. Alcohol is used as a raw material in the chemical and other industries, and it's also used to make auto fuels, but we do drink an awful lot of it. You can start your share now. Why not try something produced from the fermented juice of grapes? Wine making is now a global industry, although France and Italy still account for half the world's production. The classic wine grape has the scientific name *Vitis vinifera.* The important cultivars include Sauvignon (red Bordeaux); Pinot Noir (the main red Burgundy grape); Riesling and Silvaner (for German white wines); Barbera and Freisa (northern Italian wines); and Palomero, the main sherry grape. The whole of the black grape is crushed to make red wine; it's the grape skin pigment that makes it red. Black or white grapes can be used to make white wines, but only the pressed juice is used, and extraction of skin pigments is avoided. Of course, the quality of the wine depends on the grape used (and what the growing season was like that produced it), on production techniques, and on fine points like the soil type and whether the vineyard is on the north slope or a south-facing slope, and the color of the owner's socks.

Experts really can identify all these features from the taste and aroma of the final product. (OK, I admit it, not *all*, I was lying about the socks). The crucial feature is the controlled fermentation by the yeast, *S. ellipsoideus*, which may come from natural sources (the grapes or the preparation machinery) or from a starter yeast culture. After the yeast fermentation, quality wines take one to four years to age in wooden casks. For some wines, a bacterial fermentation is encouraged during aging to mellow the taste by reducing acidity. To make sparkling wines, sugar, a little tannin, and a special strain of *Saccharomyces ellipsoideus* that can form a granular sediment are added when the wine is bottled. A secondary fermentation in the bottle produces carbon dioxide and results in formation of an unstable compound with alcohol (called ethyl pyrocarbonatene), which gives the characteristic lingering sparkle of naturally produced sparkling wines (compared with the brief sparkle of those that have only had carbon dioxide injected into them under pressure.

You might like a glass of a fortified wine (one which has had up to 20 percent brandy added to it) as an aperitif or even at the end of your meal. Sherry is a fortified wine, made from a particular grape and a secondary growth of yeasts in the maturing vats, which creates the characteristic sherry flavor compounds. Vermouths are wines flavored with herbs, like wormwood, and with extra alcohol added. Port is a red wine in which the primary fermentation was stopped by adding alcohol or brandy while some sugar still remained. The special flavor of Madeira results from a heat treatment of the fermented wine before extra alcohol is added.

Personally, I like wine, particularly red wine, well enough, but I do prefer a glass of beer to start a leisurely meal. All ales, beers, and lagers are made from malted barley. The process starts by encouraging barley grains to germinate. In two to four days the sprouting seeds start the digestion of their stored starch, producing more soluble sugars. Then the sprouted grain is killed by slow heating and mashed into hot water with other cereals like maize, wheat, or rice. Finally, the sweet mix is boiled with hops to add bitter flavors to the beer. The boiled liquid is cooled and passed to the fermentation vessels for fermenting. After fermentation, the beer is stored (conditioned) to remove harsh flavors. Ale is conditioned for a few weeks, but lagers are left at close to 0 degrees for several months before being filtered, carbonated, and bottled.

While we sip our drinks and look at the menu, the waiter will bring us some bread rolls. Leavened bread is another product of the fermentation of sugars from cereal grains by *S. cerevisiae*. Its structure also depends on the gooey-gluey properties of the wheat gluten protein. It's another easy technology that has been with us for thousands of years. Mix wheat flour with water, yeast, salt, sugar, and some fat, and the dough will stick to anything; it's the gluten that does the sticking. Put the dough in a warm place for at least an hour, ideally two, and the yeast produces carbon dioxide and alcohol. The carbon dioxide gas is trapped into bubbles by the gluten and as more and more bubbles are formed in the dough they make it rise. After this fermentation period, the dough is cooked, the alcohol evaporates, and the bubbly structure of the dough is turned into the open structure of bread.

OK, so let's look at that menu. I think fungi will feature prominently in my choices, but I guess you expected that! Fungi are an ideal food because they have a fairly high content of protein (typically 20 to 30 percent crude protein as a percentage of dry weight), and they contain all the amino acids that are essential to human health, and several of the vitamins, too. Fungal tissue is easily digested, and the walls of fungal cells provide a good source of dietary fiber. Possibly the most important attribute of all fungal food, though, is that it is virtually free of cholesterol. Fungi just don't use cholesterol in their membranes to the extent that animals do (fungi use a different sterol that doesn't accumulate in humans). Consequently, fungal foods compare very favorably with meats on health grounds and on grounds of nutritional value. I believe they also compare well on grounds of taste. So I think I'll go for a "fruits of the forest" style mushroom starter, just a selection of wild mushrooms lightly tossed in a hot pan with melted butter and a pinch of sugar and served with fresh crusty bread and a wedge of lemon.

Collecting mushrooms for food is an age-old tradition, on a par with collection of berries and other forest fruits. In several forests in the United States and Europe, commercial mushroom picking has become big business. Chanterelles, morels, and truffles probably represent the best expression of this in the popular imagination because of the history and mystique associated with the industries in Europe, particularly France and Italy. Their histories include festivals and markets that associate folk events and heighten interest in and appreciation of the qualities of the products themselves. Those qualities contribute to the mystique, prompting discussion of how tasty and flavorful competing collections might be. The truffle probably has more mystique than most because this subterranean delicacy is still hard to find and harvest, so there is the added mystery of detecting the presence of a truffle 10 or 20 centimeters underground with the help of forest lore, flies, or trained pigs and dogs. There are about 70 different truffle species. The most highly prized in French cuisine is the black truffle of the Périgord, *Tuber melanosporum*, but in Italy, the white truffle of Alba is considered the true delicacy. Deciding these matters creates lots of fun and innocent enjoyment, and it makes lots of money. The world market for chanterelles (collected, not cultivated) was estimated recently at more 1.5 billion U.S. dollars. Add the value of tourism and peripheral matters like cookery programs on TV, recipe books, and magazine articles, and the collection and appreciation of these fungi becomes a very big industry indeed. The demand for wild mushrooms has grown sharply since about the early 1980s. Specialty mushrooms have always been harvested and shipped to distant markets. There is a story that as early as 1872 New Zealand had a fungus-based industry earning, eventually, hundreds of thousands of pounds annually (at nineteenth-century values!) by collecting the wood ear fungus for sale in China. The then Colonial Secretary of Hong Kong, in answer to an inquiry from the New Zealand Colonial Secretary in 1871, reported that the fungus was used as a medicine "much prized by the Chinese community." The local entrepreneur, a man named Chong, paid colonial farmers 4 pence per pound weight of sun-dried fungus and is said to have purchased an average amount on each market day worth 65 pounds sterling. At the 4 pence per pound rate, Mr. Chong was collecting just less than two metric tons each market day! This trade was killed off when commercial cultivation of the fungus made collection from nature uncompetitive. This story emphasizes the rural ideal of collection of a cash crop by local residents for sale elsewhere: generally small and scattered enterprises in which pickers sell and ship most of their own harvest. The trade may be seasonal, and the volume of mushrooms picked may be relatively small.

All of this changed during the late 1980s, mainly in the amount and manner in which mushrooms were harvested, sold, and handled. The commercial picking industry has now expanded to a system of harvesters, buyers, processors, and brokers. Harvesters locate and pick the mushrooms. Buyers, typically associated with a specific processor, set up buying stations near wooded areas known to produce mushrooms and advertise their willingness to buy. Processors grade, clean, pack, and ship the product and provide the cash directly to the field workers. Brokers market the mushrooms around the world. This is a model that has become common in Europe and the United States. One of the things that makes it viable is the easy access to rapid transcontinental and intercontinental transport. As transport and communications continue to improve, the commercial picking industry is bound to continue to expand. Even remote areas may experience commercial picking of especially valuable species. Apparently, helicopters are routinely used to transport matsutake collected (mostly for export to Japan) in roadless areas of interior British Columbia. There is significant international competition, and international markets and prices can fluctuate wildly from year to year, and even within a season, as global weather patterns produce good or poor crops in various locations.

When prices are high, large numbers of pickers may congregate in small areas and then the operation is seen as a problem, although the exact perception of what the problem consists of depends very much on the standpoint of the observer. The British *Daily Telegraph* newspaper had a report on July 1, 1993, headlined "Mushroom Rustlers Shoot It Out" that told of heavily armed mushroom rustlers taking part in mushroom wars in Oregon's forests as they battle for rare mushrooms. A Morrow County Sheriff's Deputy was quoted as saying that "nearly all the mushroom pickers are armed and it's real scary." Apparently, two people had died in the shootings, and several others had been mugged for their mushrooms. Now, that's a problem! But it's not a problem with the mushrooms.

In any one region there may be thousands of pickers harvesting fungi for commercial purposes from both private and public lands. The knowledge we have about the ecology of wild edible mushrooms is incomplete, and this ignorance is at the center of the (unarmed) conflicts that are arising among commercial pickers, conservationists, and local residents. The three parties do not always align as might be expected. A successful commercial picking job can completely denude a region of marketable mushrooms in just a few hours. Local residents see this as destruction of a natural resource that belongs to the people and expect the support of conservation-minded mycologists in the campaigns that result. Unfortunately, ownership of the resource is not always entirely transparent. The people may be allowed to enjoy a woodland for what it is by a generous landowner who subsequently is enlightened to the cash value of his mushroom crop. Similarly, it is not at all clear that picking mushrooms does any damage that a conservationist would be concerned about. In the United Kingdom, the activities of commercial pickers have been likened to the activities of factory fishing boats, which take fish of all ages and consequently damage the breeding stock of the fishery to the point where the fish population declines drastically. This is an emotive comparison for the United Kingdom, which has experienced a catastrophic decline in its own fishing industry, but it has almost no biological relevance to commercial mushroom picking. Mushrooms are not individuals, but simply the fruiting structures arising on underground fungal growth, more like apples on a tree than fish in the sea. Removing one generation of fruit bodies will probably *encourage* a new generation to emerge; this is similar to pruning. Certainly, continued productivity of mushroom farms is enhanced by regular harvesting. Try to explain that to irate residents who have just seen "their" mushrooms disappear down the road on the back of a truck, and their respect for academic mycologists evaporates!

The growth in commercial mushroom picking of recent years is unlikely to have generated so much concern had it not coincided with the growing debate about conservation and the damage done to natural biodiversity by commercial pressures. Revelations of the extent of commercial picking, or indeed the arrival of commercial pickers in a woodland, fuel concern that both timber and mushroom harvesting adversely affect the sustainability of wild mushroom populations. You don't have to be an expert to recognize that the woods and forests are being changed by commercial pressures of all sorts; from pressure to use the land for other purposes to pressure to grow a more profitable tree crop. Couple this with the known adverse effects of atmospheric pollution (especially severe in northern Europe), and it's easy to see how commercial mushroom picking might be viewed as one step too far.

Some U.S. agencies are restricting mushroom harvest in particular forest areas because of these uncertainties, and additional regulation and legislation will no doubt be called for. Data from Europe, however, has not blamed decline in populations of fungi in the last 30 years to collection of mushrooms by commercial pickers. Rather, alterations in forest habitats by agriculture and urban development has led to changes in fungus composition. Importantly, though, decline in mushroom populations over these years has outpaced the loss or alteration of habitat, changes in forest age, change in tree composition, and change in forest structure. A small change to the forest can cause a big change in the fungi. The key response to all this seems to be effective management of the forest resource as a whole, fungi included. It may make more (economic) sense to cultivate trees for the sake of the fungi with which they are associated. The forest then becomes the mushroom farm, a sustainable resource that the public can enjoy while the mushroom harvester profits from it. Truffle cultivation is a successful model of what might be done. The truffle is the underground fruit body of a mycorrhiza of the oak. Truffle cultivation was first achieved early in the nineteenth century when it was found that when seedlings adjacent to truffle-producing trees were transplanted, they too began producing truffles in their new location. *Truffières* or *truffle groves* have been established throughout France in the past 100 years, and the value of the crop is such that the practice is now extending around the world. Truffières are started by planting oak seedlings in areas known to be infested with truffle fungi. The truffles begin to appear under such trees 7 to 15 years after planting, and cropping will continue for 20 to 30 years. Most plants infected with the black truffle are now raised in greenhouses, although pure cultures of this truffle cannot be used yet to inoculate the roots of oak seedlings. Recently, methods have been developed to colonize plant roots with one of the white truffles, encouraging the hope that the same might be done with other truffle species.

Conventional mushroom cultivation produces a total crop of around 5 million metric tons each year. At averaged-out prices this has a retail value of about 50 billion U.S. dollars. In the mid-1970s, the button mushroom (called *Agaricus*) accounted for over 70 percent of total global mushroom production. Today, it accounts for something closer to 30 percent, even though production tonnage has more than doubled in the intervening years. The biggest change during the last quarter of the twentieth century has been the increasing interest shown in a wider variety of mushrooms. Even in the most conservative of markets (like the United Kingdom), so-called exotic mushrooms have now penetrated the market, and supplies of fresh shiitake (*Lentinula*) and oyster mushroom (*Pleurotus*) are routinely to be found alongside *Agaricus* in local supermarkets. Most of these mushrooms are cultivated fairly close to the point of sale. For example, most U.K. mushrooms originate locally or from the Netherlands or Ireland. The industry is truly international, however, and a small supermarket local to my home in south Manchester regularly displays punnets of fresh enoki (*Flammulina*), which are grown in Chile. This indicates that intercontinental air transport makes the 12,000-kilometer distance irrelevant, and that the production costs are sufficiently low to enable reasonable pricing in such a distant market.

Mushroom cultivation is the next-biggest biotechnology industry after alcohol production, and the mushroom industries of the world all depend on some form of solid-state fermentation. In the European tradition this has come to mean cultivation of a mushroom crop on compost. Similar approaches were developed for oyster and paddy straw mushrooms in the Orient, though in the Chinese tradition the typical approach is to cultivate the crop of choice (*Lentinula*) on wood logs.

Good compost is the essential prerequisite for successful farming of the *Agaricus* mushroom, but compost preparation is a smelly process, and even the most modern installations have a severe impact on their neighbors. The basic raw material for mushroom compost in Europe is wheat straw, although straws of other cereals are sometimes used. Ideally, the straw is obtained after it has been used as stable bedding and is already mixed with horse manure. On a commercial scale, this is not possible, so other animal wastes, like chicken manure, are mixed with the straw, together with gypsum and large quantities of water. The European mushroom industry is said to have originated in the caves beneath Paris at the end of the nineteenth century. It probably emerged from the food provisioning functions of the kitchen gardens on the estates of the European aristocracy. Some of the surviving records of such estates refer to manured and composted plots set aside for mushroom production. The compost used, and its preparation, would very definitely be familiar to the very competent gardeners of the day. The current industry depends on a compost that is very selective for the crop species. Although widely distributed in nature, the *Agaricus bisporus* fungus is rarely encountered because it produces relatively few mushrooms in the wild, and only infrequently. The industry we know today seems, therefore, to be the result of a remarkable joint evolution during which an otherwise ordinary horticultural compost was developed that achieves high cropping densities with an otherwise unremarkable and not very abundant mushroom. And all without a genetic engineer in sight!

The beginnings of mushroom fruit bodies (so-called pins or pinheads), which are more or less spherical and have a smooth surface, will be seen about three to four weeks after the compost is first seeded (or the actual term, *spawned*), and about one to two weeks later marketable mushrooms can be harvested. Successions of mushrooms then develop in a series of flushes about eight days apart, each taking about five days to clear from the beds. Growers expect to harvest between three and five flushes from each spawning cycle, with a 25-kilogram total yield from every square meter of growing tray. After the final pick (7 to 10 weeks after spawning) the compost is spent, and the cropping room is emptied, cleaned, sterilized, and filled with the next crop. On most farms a new crop is filled every one or two weeks throughout the year. So a mushroom farmer is likely to see more crops in one year than a cereal farmer will see in a lifetime!

It's different in China, of course. The main crop is the black-oak or shiitake mushroom (official name *Lentinula edodes*), which is traditionally grown on deciduous hardwood logs (oak, chestnut, hornbeam) and is still very widely grown like this in the central highlands of China. To put this statement into perspective, the most frequently used method in China is the traditional log-pile approach; the growing region covers an area about equal to the entire land area of the European Union. The logs considered suitable for shiitake production are about 30 centimeters in diameter and 1.5- to 2-meters long, and are normally cut in spring or autumn of each year to minimize preinfestation by wild fungi or insects. Holes drilled in the logs (or saw- or axe-cuts) are packed with spawn, and the spawn-filled hole is then sealed with wax or other sealant to protect the spawn from weather, insects, and competitor fungi. The logs are stacked in laying yards on the open hillside in arrangements that permit good air circulation and easy drainage and warm temperatures (24 to 28°C). The logs remain here for the five to eight months it takes for the fungus to grow completely through the log. Finally, the logs are transferred to the raising yard to promote fruit body formation. This is usually done in winter to ensure the lower temperature (around 12°C) and increased moisture that is required for fruit bodies to start. The first crops of mushrooms appear in the first spring after being moved to the raising yard. Each log will produce about two kilograms of mushrooms, each spring and autumn, for five to seven years. This traditional approach to shiitake production is expensive and demanding in its use of both land and trees. Some commentators estimate that there are 10 million mushroom farmers in China; if this is true, the traditional use of locally cut logs is likely to devastate the hill forests. This is a good reason why more industrial approaches are being applied to shiitake growing. Hardwood chips and sawdust packed into polythene bags as artificial logs provide a highly productive alternative to the traditional technique, and the cultivation can be done in houses (which may only be plastic-covered enclosures) in which climate control allows year-round production.

The Chinese straw mushroom *(Volvariella volvacea)* is grown mainly on rice straw, although several other agricultural wastes make suitable substrates. Preparation of the substrate is limited to tying the straw into bundles that are soaked in water for 24 to 48 hours. The soaked straw is piled into heaps about 1-meter high, which are inoculated with spent straw from a previous crop. An important reason for the remarkable increases seen in production of certain mushrooms has been the use of substrates that are waste products from other industries. For example, although the Chinese straw mushroom is traditionally grown in Southeast Asia on rice straw, it can be grown on cotton waste. Cotton waste gives higher yields and is also more widely available than rice straw, so it is a far cheaper substrate (the higher cost of rice straw does not indicate any intrinsic value but rather the price of transporting it to a non-rice-growing region). Oyster mushrooms are also easily grown on a wide range of agricultural wastes.

There are quite a few other cultivated mushrooms, something approaching 15 or more, but we have quite enough for a starter already, and its time to think about a main course. Main course, yes that usually means a large lump of protein like a steak or something else that used to run around going moo, bah, oink, or cluck, but it doesn't have to be meat. Have you considered single-cell protein? Understandably, it doesn't appear under that name on the menu, and I've got to admit that the emphasis given between the 1950s to the 1970s to the production of single-cell protein is now almost forgotten. In those days the hope was that microbes could provide a means of solving the world's food shortage by industrial production of cheap protein alternatives to meat protein. Several fortunes were invested in the idea and were duly lost when it turned out that the people who were short of food didn't want to eat industrially produced single-cell protein, thank you very much. Solution of problems on that scale has more to do with politics and economics than with biotechnology. Today, the emphasis has moved toward the use of single-cell protein for animal feed, and the only successful fungal product currently on the market is the mycoprotein Quorn. *Myco-protein* is the term coined by the United Kingdom Government's Foods Standards Committee to serve as the general name for a food product resulting from the growth of a selected strain of the filamentous soil-fungus *Fusarium*. This is grown in a very large (45-meter tall) air-lift fermenter. This is a high-technology product that took a long time (and a lot of money) to develop and even

Eating fungi is fun and healthy. Fungi contains a good amount of vitamins, protein, fiber, and no cholesterol! You could choose to eat pieces or slices of Quorn (top images) or take the natural option and cook yourself a cheap and easy meal (bottom photographs).

longer to get official approval to use it as human food. Quorn is marketed as "The tasty, healthy, alternative to meat" because its filamentous structure enables it to be processed to simulate the fibrous nature of meat. Coupled with the inherent nutritional value of fungal tissue, this permits the product to be sold as a low-fat, low-calorie, cholesterol-free health food to consumers who can afford to choose Quorn as a meat substitute. The retail prices of Quorn exceed those of most meats and are two to three times greater than retail prices of mushrooms. Quorn mince and Quorn pieces can be used in cooking recipes in much the same way as meat products. If you talk nicely to the chef I'm sure he'll be delighted to prepare you a choice Quorn burger or Quorn fillet steak.

Me? No thanks, I'll write about it but I don't need to eat it. I'll settle for a mushroom stroganoff. Why eat substitute meat when you can eat real mushrooms for half the price? You can feed four with this recipe: heat about 30 milliliters (1 ounce) of good olive oil in a large frying pan with 350 grams (11 ounces) of chopped onion. Cook until the onion goes translucent (about five minutes) then add 700 grams (22 ounces) of chopped mushrooms (assorted, equal amounts of oyster, shiitake, and brown-cap *Agaricus* mushrooms are recommended, or all of one sort according to your preferences) with a crushed clove of garlic or two (to taste), and cook over a moderate heat for 5 to 10 minutes. The oyster mushroom stems will take longest to cook; if a fork will go through them easily then they're done. Stir in 300 milliliters (10 ounces) of a stock made up from stock cubes (vegetable if you want a vegetarian dish, chicken otherwise) together with 15 to 20 grams (about .5 ounce) of cornstarch as a thickening agent. Bring to a boil and simmer for 5 to 10 minutes. Remove from the heat and season with pepper and salt, and lemon juice and/or tabasco sauce to taste. Finally, stir in 300 milliliters (10 ounces) of fromage frais (you can use sour cream if you don't care about either your waistline or the cholesterol) and heat through until piping hot, but do not allow to boil. Serve immediately onto a bed of rice or pasta. That goes down well with a nice red wine from Spain!

As well as being used directly as food, fungi are also used in the processing of various food products. In these the fungus is mainly responsible for the production of some characteristic odor, flavor, or texture, and may or may not become part of the final edible product. Growing fungi on water-soaked seeds of plants is a popular way to produce several foods in Asia, including soy sauce and various other fermented foods. In soy sauce production, soybeans are soaked, cooked, mashed and fermented with two molds called *Aspergillus oryzae* and *A. sojae*. Depending on the size of the factory, the soybeans may be fermented in fist-sized balls (the traditional method) or on trays. When the soybean substrate has become overgrown with the fungus, the material is mixed with salt and water, and the fermentation is completed in the brine. The biggest industrial units today use a continuous process in which defatted soybean flakes, moistened and steam-sterilized are mixed with ground, roasted wheat. The mixture is turned mechanically to ensure even growth of the two molds for two to three days; then it is transferred to brine and inoculated with a bacterium and a yeast. The brine fermentation takes six to nine months to complete, after which the soy sauce is pressure-filtered, pasteurized, and bottled.

As an example of a fermented food, rather than a flavoring condiment like soy sauce, the Indonesians make tempeh: a white cake produced by fermentation of partially cooked germinated soybeans with a different mold (*Rhizopus oligosporus*). The fungus binds the soybean mass into a protein-rich cake that can be used as a meat substitute. This is being increasingly widely sold in the vegetarian market. There are a variety of other fermented products of this sort. Ang-kak is a rice product popular in China and the Philippines, which is fermented using a fungus called *Monascus purpureus*. This fungus produces red pigments as well as alcohol which are used for red rice wine and food coloring.

These exotic products have their place, but cheese is a very good way of ending a meal. If you look in the dictionary you'll find that cheese is a solid or semisolid protein food product manufactured from milk. Before the advent of modern methods of food processing, like refrigeration, pasteurization, and canning, cheese manufacture was the only method of preserving milk. Although basic cheese making is a bacterial fermentation, there are two important processes to which fungi contribute; these are the provision of enzymes for milk coagulation and mold-ripening. Cheese production relies on the action of enzymes that coagulate the proteins in milk, forming solid curds (from which the cheese is made) and liquid whey. Traditional cheese-making uses animal enzymes (called chymosin or rennet) extracted from the stomach membranes of unweaned ruminants. Rapid expansion of the cheese-making industry caused attention to shift to alternative sources of such enzymes, and molds like *Aspergillus* and *Mucor* have supplied these to the extent that around 80 percent of cheesemaking now uses coagulants from non-animal sources. Recently, animal enzymes produced by genetically modified microbes have entered the market. Indeed, in 1988, chymosin was the first enzyme from a genetically modified source to gain approval for use in food manufacture. Today, about 90 percent of hard cheese production depends on enzymes from genetically modified microbes (mostly yeasts) for the coagulation step.

Mold ripening is a different matter. It is a traditional method of flavoring cheeses which has been in use for at least 2,000 years. Blue cheeses, such as Roquefort, Gorgonzola, Stilton, Danish Blue, and Blue Cheshire, all use *Penicillium roquefortii,* which is inoculated into the cheese prior to storage at controlled temperature and humidity. The fungus grows throughout the cheese, producing flavor and odor compounds. Camembert and Brie are ripened by a mold called *Penicillium camembertii,* which changes the texture of the cheese rather than its flavor. This fungus grows on the surface of the cheese, extruding enzymes that digest the cheese to a softer consistency from the outside toward the center.

You know, if we finish off the last few crumbs of cheese on that plate with just one more tiny glass of Frangelico liqueur (that hazel nut flavor complements Roquefort so well), we could go quietly to sleep in the corner until the next chapter . . . and nobody will notice. . . .

8

The Old Kingdom in Time and Space

Call me Ishmael. Well, you can call me anything, really, just don't call me a monkey; my opinion of your general knowledge and level of education will be greatly reduced if you do. I'm a hominid primate and proud of it. Monkeys have tails and questionable habits and are only very distantly related to me because their evolutionary line launched off on its own about 40 million years ago. But then, thanks to TV and printed media, we all know about that. Chances are that we've seen or read about Jane Goodall and her chimpanzees, about the plight of mountain gorillas in the interminable civil wars in Africa, and about the sad predicament of orangutans in the dwindling forests of Sumatra and Borneo. We know about Lucy, the 3-million-year-old fossil from Ethiopia, and, indeed, most people probably have quite a sophisticated understanding of animal relationships and evolution. It's likely that this is centered on relationships to and evolution of the human animal. However, there are other bits of general knowledge that are equally impressive.

In the same TV programs and the same magazines we have watched whales migrating from Arctic to Antarctic; been amazed by the aquabatic feats and communication skills of dolphins; and appreciated the family structure of killer whales (and happily chatted about Orca families called pods). So when Herman Melville has the Mate Starbuck ask of Captain Ahab ". . . what more wouldst thou have? Shall we keep chasing this murderous fish till he swamps the last man?" when talking about Moby Dick, the great White Whale, I think most people today would stop and say that there's something wrong with that "murderous fish" bit.

You might think the mistake is excusable in the context of a nineteenth-century writer putting words into the mouths of nineteenth-century fishermen. After all, the King James edition of the Bible tells that "the Lord had prepared a great fish to swallow up Jonah. And Jonah was in the Belly of the fish three days and three nights." But to think that the whale = fish equation in *Moby Dick* is an ignorant mistake would be doing Melville an injustice. Melville, through his spokesman Ishmael, explains it like this.

> There are some preliminaries to settle. First: the uncertain, unsettled condition of this science of Cetology . . . that in some quarters it still remains a moot point whether a whale be a fish. In his *System of Nature,* A.D. 1758, Linnaeus declares, "I hereby separate the whales from the fish. . . ." The grounds upon which Linnaeus would fain have banished the whales from the waters, he states as follows: "On account of their warm bilocular heart, their lungs, their movable eyelids, their hollow ears, penem intrantem feminam mammis lactantem, and finally, ex lege naturae jure meritoque." I submitted all this to my friends Simeon Macey and Charley Coffin, of Nantucket, both messmates of mine in a certain voyage, and they united in the opinion that the reasons set forth were altogether insufficient. Charley profanely hinted they were humbug. Be it known that, waiving all argument, I take the good old fashioned ground that the whale is a fish, and call upon holy Jonah to back me.

Well, we've all seen so much on TV and film, books, and magazines that most people these days would think it uneducated and unsophisticated to hear someone describe a whale as a fish. I'm sure that even Simeon Macey and, especially, Charley Coffin, of Nantucket, would be convinced. In these enlightened times, then, why do so many people think fungi are plants?

They're not plants. They never have been and they never will be plants. Fungi form a group of organisms entirely distinct from both animals and plants. Unfortunately, fungi have been damned by history to be lumped in with plants at almost every turn. Historically, the study of fungi, a science that is called mycology, has origins in two activities, both strongly associated with plants. The first emerged from the ancient habit of collecting the larger fungi, that is, mushrooms and toadstools, for food and for medicine. This has been done for thousands of years by most human societies around the world, and as they collected the mushrooms and toadstools they were also collecting berries, fruits, and plants from the same locations. And so the fungi became identified with vegetables and plants. The second activity was the scientific study of fungi. This developed in a systematic way in the nineteenth century and the first academic mycologists, aside from collecting and cataloging, were mainly interested in understanding the role of fungi in causing plant diseases. And so fungi inevitably fell within the orbit of the botanists, and mycology never developed an identity of its own. Mycologists have aided and abetted in this. They have sat comfortably in departments of botany for most of this century, although presently, unfortunately, when downsizing becomes necessary, the mycologists are the first to go. The only alternative home that mycologists seen able to find is in microbiology. The bothersome aspect of this is that the biggest organism on the planet is a fungus. *Micro*biology?

In the mid-1960s, a revolution occurred in the understanding of relationships between organisms. However, there are still a large number of people around who completed their education before this time, and who are firmly convinced that fungi are plants: peculiar plants, maybe, but plants nevertheless. This idea is now accepted as completely wrong, and plants, animals, and fungi are assigned to three quite distinct kingdoms of higher organisms. Arranging organisms into kingdoms is a matter of what is called systematics, an agreed scheme of naming things. It really is only a matter of looking, comparing, and making categories, but the arrangement we have reached now does seem to accord with current ideas about the early evolution of these organisms. So it seems to be a natural, meaningful systematics in which the fungi have been allocated a kingdom, the so-called fifth kingdom, all of their own (the other four kingdoms being bacteria, plants, animals, and primitive single-celled creatures called protists).

Ideas about categorizing things change as more information comes to light, so some people believe there should be seven, or nine, or even more kingdoms. But these are minor adjustments that still leave most of the fungi in a kingdom of their own, and there are probably more people who know it as *the fifth kingdom* than by any other number. A major aspect of the original 1960s definition of the kingdoms was their nutrition: plants use the direct radiant energy from the sun to make their food. Animals, from amoeba to killer whales, engulf their food. And fungi? Well, fungi degrade food externally and absorb the nutrients which are released. Once this apparently simple basis for making the grand separation between kingdoms has been used, numerous other differences in structure and lifestyle become evident. One thing worth emphasizing is that the three kingdoms are very different from one another in ways that are crucial to determining shape and form. A key feature during the embryology of even lower animals is the movement of cells and cell populations, so cell migration (and everything that controls it) plays a central role in animal development. Being encased in walls, plant cells have little scope for movement, and their changes in shape and form are achieved by regulating the orientation and position of the wall that forms when a plant cell divides. Fungi are also encased in walls, but their basic structural unit is a tubular, thread-like cell called a hypha. The hypha has two peculiarities that result in fungal development being totally different from that in plants: it grows only at its tip and new walls form only at right angles to the growth axis of the hypha.

This is a nice bit of fine detail that pleases the academic in me, but it has a significance beyond that because it raises an important consequence for arguments about the evolution of these organisms. The evolutionary separation between the major kingdoms must have occurred at a stage when the most highly evolved things were single cells. The consequence is that each kingdom must have learned independently how to organize populations of cells in order to make the multicellular organisms we now know as mushrooms, mice, or marigolds. So studying how a mushroom makes a mushroom is an investigation every bit as deep and meaningful (and difficult) as studying how a human embryo develops or how a tree is shaped and sculptured in the forest. Sadly, in the popular imagination, mushrooms don't have the same status as human animals or forest trees.

The man who is usually thought of as the "Father of Mycology", Elias Fries, and his contemporaries in the early nineteenth century put far too much emphasis on shape in their attempt to understand mushrooms, toadstools, and their relatives. This fundamental mistake is still with us. Academic mycologists may have moved well away from this position now, but they have taken a long time to do so, and external shape is still the guiding light in the minds of many nonacademic mycologists. With plants and animals, shape was important in ancient herbals and bestiaries, but it has not been a dominant factor in twentieth-century science. Indeed, as my quotation from *Moby Dick* shows, the "Father of Classification", Linnaeus, distinguished the shape of whales from the shape of fish on the basis of comparisons of their structure, anatomy, and development. This is not so for fungi. Most of the early workers were content to rely solely on current shape and form, the most simplistic sort of study. Detailed studies of development, structure, and anatomy did not start until the middle of the twentieth century, and even today their practitioners have to fight to be heard, and deep ignorance abounds.

Remember that classification of organisms is an exercise in arranging them into groups to make it easy to study them and, more importantly, to make it easy to interpret the results of the studies. It's a clerical exercise, like arranging all the papers in an office in files and then into filing cabinets. The arrangement that Fries produced used the shape and form of the fruit body and especially the nature of the tissue on which the spores were made. So he had a group called *agarics,* which have plates (or gills) beneath an umbrella-shaped cap, just like the ordinary cultivated mushroom. And mushrooms with gills were filed into that group, regardless of their other characters. This agaric group was contrasted with fruit bodies that had tubes (or pores) in a spongy layer beneath the cap (called *polypores*). Toadstools in this group were called *boletes,* but *bracket fungi,* whose fruit bodies grow directly on the trunks of trees, were also included. Then there were those with teeth or spines hanging down below their cap or bracket, and these were called *hydnoids.* Other major groups included some with spores formed over the outside of a club-shaped (called *clavarioid*) or coral-like (called *coralloid*) fruit body, and then there were the completely enclosed fruit bodies (called *gasteroid*) that had their spores inside the fruit body like puff-balls.

On the face of it, this is a nice simple scheme. Just the sort of thing you could apply as you trek through a forest. Sadly it was applied too rigidly. It was as if the so-called experts insisted that everything that flew should be called *bird* whether they were actually fruit bats, bumblebees, or buzzards. When examined closely, one can see the limitations of the scheme, but very few mycologists looked closely enough. Even though some studies done soon after Fries published his work indicated that his groupings were artificial, his views were so fervently believed that at the end of the nineteenth century such suggestions were considered heresies.

In fact it's taken over a hundred years to break the stultifying grip of the Friesian system of classification of larger fungi. The biggest changes have come from work done in the last 25 years of the twentieth century. Work that has used developmental features and detailed comparisons of anatomy, chemistry, and microscopic characters to reveal natural groupings and evolutionary relationships. Ironically, the zoologists battled through this sort of argument in the development of evolutionary ideas in the first quarter of the twentieth century. Initially, animal evolution was thought of as resulting mainly from modification of adult form, and development was seen as a recapitulation of previous mature stages. This was encapsulated in "the individual in its development recapitulates the development of the race" in MacBride's *Textbook of Embryology* in 1914. Progressive views were diametrically opposed, and by the mid-1920s it was widely agreed that animal embryology does not recapitulate evolution, but contributes to it. Reproduction, or rather *success* in reproduction, is the winning ticket in the evolutionary lottery. So all features that influence reproductive success are subject to evolutionary selection. For an animal, that might mean efficiency in finding a mate; efficiency in egg-laying; efficiency in dispersal of larvae; efficiency in care of the young; or any of a host of other factors that contribute to one set of genes being distributed to the next generation in preference to some other competing set of genes. If the cosmic watchmaker is blind, then that set of evolutionary principles apply just as much to fungi as to animals and plants. Only the details change.

The function of the fungal fruit body is to distribute as large a number of spores as the structure will allow. The familiar mushroom shape has evolved to give protection to the developing spores. It really is an umbrella protecting the spores from rain. The first step in improving the basic mushroom shape is to expand spore-production capacity. Making gills (plate-like downward extensions of the cap) and pores (tubular excavations into the cap) are both strategies to increase the surface area available for spore production. If that has positive evolutionary advantage, is it any wonder that careful observation of developing fruit bodies shows that there are at least 10 different ways by which the mushroom shape can be constructed? It's relatively easy to show that geometrical constraints make pores a less efficient way of expanding spore production than gills. So, don't be surprised to find that there are some gilled mushrooms that are closely related to polypores and only distantly related to real agarics. Oyster mushrooms and the shiitake mushroom are like this. Presumably, at some stage in their evolution they found advantage in folding their spore-forming tissue into gills and have now converged onto the agaric shape. It's *convergent evolution;* swimming mammals evolved toward efficient swimming in water and ended up fish-shaped, like dolphins, because fish also but quite independently evolved toward efficient swimming in water.

Look closely at the tissue structure for more wonders of evolutionary architecture. There are different ways of constructing gills. All mushrooms must increase in size as they develop. Some hold the number of cells unchanged but pump fluids into these cells to increase their size by 10, 20, or 50 times. Others hold the size of cells unchanged and just make more of them to increase the volume of the tissue. Both strategies, though, seem to use the same simple management system whereby one cell organizes and controls a rosette of cells immediately surrounding it. These little families of hyphae orchestrated by a central inducer hypha are called *Reijnders' knots* after the man who first described them.

Look closely at the lifestyle strategies of fungi, and you will find some interesting behavior patterns. Bracket fungi achieve massive spore production by increasing the lifetime of the fruit body. At the extreme, the fruit body is adapted to be perennial. These fruit bodies are often described as woody, but obviously, because fungi are not plants, they cannot use the plant-like wood components, and in another example of convergent evolution, they have developed their own solutions to the same challenges solved by woody plants. If fungi are to last several growing seasons, they need mechanically strong structures that must be resistant to attacks by pests and microbes and to adverse weather conditions.

At the other extreme of the lifetime strategy spectrum, there are some fungi whose fruit bodies last barely more than a day. They may mature overnight and be dead and gone by the next night. These mushrooms are stripped down for athletic action. They tend to be small and delicate and adapted to get maximum spore yield from minimum mass of fruit body.

Stink horns are interesting because they parallel mushrooms in gross morphology; they have recognizable caps and stems. However, the whole structure is adapted to insect dispersal as opposed to wind dispersal of the spores. Like insect-attracting flowering plants, these fungal fruit bodies sport bizarre shapes, colors, and penetrating smells to attract flies and other

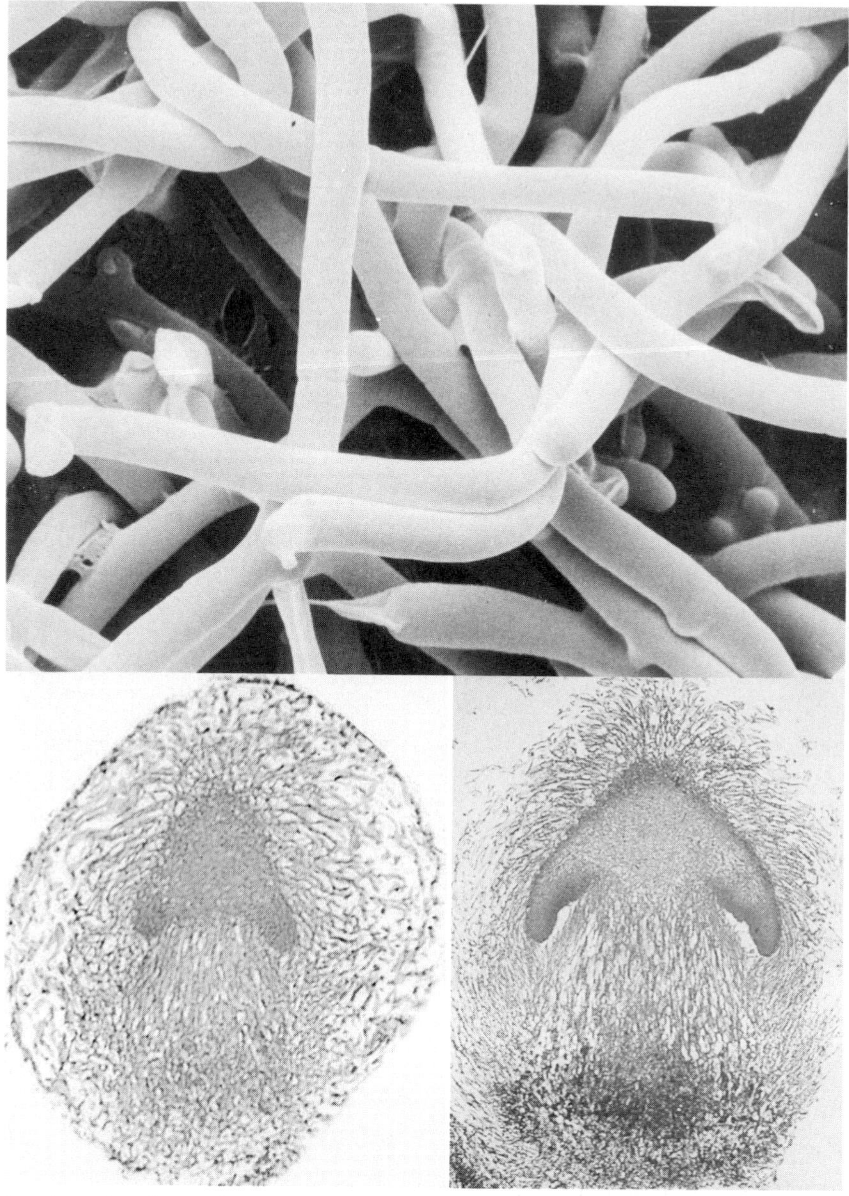

Fungi produce long cylindrical cells called hyphae, shown greatly enlarged in the electron micrograph (of hyphae of the mushroom *Pleurotus*) (top). This growth pattern does not reduce their ability to make complex tissues. A young mushroom is an embryo. The microscope images at the bottom show vertical sections of mushroom fruit bodies less than one millimeter tall (left) and just over that size (right), only 1 percent of the final size of the mushroom.

insects. A distinctive odor is also important for fruit bodies formed under the soil surface. Truffles are the obvious examples, but other fungi use the same strategy of attracting animals (some produce an odor resembling the male sex hormone of pigs) to dig them up and help disperse their spores.

Fungal systematics is now coming of age. Almost a century behind its sister sciences of botany and zoology, mycology is now appreciating that fruit-body shape should not hold such a central position as it once did. This is a matter of interpretation of the value of particular features in establishing relationships between groups of organisms. One reason why the belief is growing that fruit-body shape is less useful in fungi is that it's become clear that it is a more flexible feature than previously thought. Variation in shape and form of mushrooms occurs at different levels and for different reasons. Many mutants or variants in shape have been induced in the laboratory or isolated from nature. These mutants have been very instructive in establishing developmental pathways, and future molecular analysis will be even more revealing. But there are also several instances where, for some reason, the development of a normal fruit body becomes disturbed without change to its genes. This sort of variability seems to be a strategy to cope with environmental stress. If there is a prime rule in fungal fruit-body development, then it seems to be *distribute spores*. If the environmental conditions are so damaging that only a monstrous distorted fruit body can be produced, then it is counted as a success if it disperses some spores. The fungi really take this tactic to extremes. An agaric fungus that normally produces a mushroom fruit body can, if the atmosphere gets too dry or the temperature too high, form coralloid or gasteroid fruit bodies alongside the normal agaric fruit bodies, which both function as spore producers and dispersal structures. To compare this with an approximate animal counterpart, the parallel would be for cats to be able, quite normally, to give birth to litters containing an odd kitten that looked like an aardvark, dog, or even iguana.

Abnormal mushrooms like this indicate that normal fruit-body development is made up of separate routines, each of which controls the structure and shape of a different component of the fruit body. So development of fungal fruit bodies in general depends on organized expression of these routines. The sequence and location in which routines are put into effect determines the shape of the fruiting structure and the progress of its development. It's similar to the way different component production lines contribute to the manufacture of a product, such as a car. Change the way the components are assembled and you change the product. In fungi, essentially the same routines can give rise to very different fruit-body shapes, depending on other circumstances.

For example, the agaric gill routine is expressed with the rule "where there is space, make gills." When this is combined with a routine providing mechanical anchorages, the gills are stretched along the lines of mechanical stress as the fruit body expands and end up straight and radial. If the anchorages are left out (e.g., because of some mutation), then the gills are formed but are never pulled straight and remain tangled and contorted.

A highly flexible developmental process like this allows the fungus to adapt to a wider range of conditions. The criterion for successful adaptation is successful production of spores, and even the most monstrously abnormal mushroom can do that. This is not true for animals and plants, where even mild abnormalities can reduce their ability to reproduce quite drastically. Fungi differ from animals and plants, therefore, by having much less selection pressure against developmental abnormality.

Development of a structure like a mushroom fruit body, flowering plant, or furry little animal involves individual cells undergoing different sorts of specialization to carry out different functions in the final structure. Generally speaking, this sort of cell differentiation involves successive steps that steadily reduce the options the cell can follow. Eventually, the cell has only one option: it is fully specialized for a particular function. Early in this differentiation pathway, the cell retains the ability to revert back to the unspecialized "embryonic" state, but as it progresses through its differentiation pathway, it becomes committed to that pathway and can turn back no more, which is another respect in which fungi differ from animals and plants. The only committed cells we've been able to find in mushrooms are those that make spores. All other cells, no matter how differentiated they become to particular functions, are able to revert to the simple original state if they are removed from the fruit body and put onto some nutritive artificial medium.

This is another evolutionary adaptation that permits flexibility. It allows the fungus to start over again if conditions really turn so nasty that continued development of the fruit body is not feasible. But this feeble grasp on their specialization also tells us something else unexpected about fungal developmental biology. Because undisturbed cells in the fruit body do not revert to hyphal growth, their differentiated state is somehow continually reinforced while they are inside the fruit body. Rather than rigidly following a prescribed sequence of steps, developmental pathways in fungi allow application of rules that allow great variability in expression. A sort of fuzzy logic in which decisions between possible pathways are made with a degree of uncertainty, being based on balancing probabilities rather than all-or-none switches.

Fungal cell differentiation is no less sophisticated or complex than is found in animals and plants, but it is very different. Fungi can vary the timing, extent, and mode of differentiation in response to external signals. They can swap growth forms and procreative phases of their life cycle. It all contributes to making them supremely able to adapt to challenging conditions. This results in a flexibility which surpasses that of other organisms and provides the mycologist with an enormous intellectual challenge.

Nutrition has always been a major characteristic in schemes of classification. Photosynthetic plants have always been clearly distinguishable from animals, which ingest, and recognizing the fundamentally different sort of nutrition fungi employ was one of the features that placed them in an entirely separate kingdom fungi. One of the original articles on this topic emphasized this point as follows: "[The] nutritive mode and way of life of the fungi differ from those of the plants. . . . Fungi characteristically live embedded in a food source or medium, in many cases excreting enzymes for external digestion, but in all cases feeding by absorption of organic food from the medium. Their organization . . . is adapted to this mode of nutrition."

Fungi have evolved to grow effectively embedded in a substrate that they digest by the excretion of appropriate enzymes. The smaller molecules produced by the activity of those enzymes are the nutrients that can be absorbed by the fungal cells. Among higher organisms, only the fungi must digest their food externally prior to absorption of the smaller molecules of which the food is composed (though this is a character they share with bacteria). There are ecological and structural, and biochemical, consequences of this. An animal, even the simplest one, can capture morsels of food by engulfing them. The food then immediately goes inside the body where it can be digested without fear of loss to competing organisms. A fungus may be capable of digesting the same food but must perform most of that digestion externally with the valuable results of the digestion being open to absorption by competitors until they can be taken in by the fungus. This may have influenced the evolution of cell walls and digestive enzymes in ways that enable the fungus to improve its competitive effort by controlling the environment in the immediate vicinity of the fungal surface.

The evolution of fungi cannot be established from a good collection of fossils. There are some fossils, but they are relatively few and scattered across evolutionary time. This has meant that we have to use evolutionary trees constructed from analysis of molecular (protein and nucleic acid) structures. This approach has been validated by comparisons in organisms (like many animal groups) for which a good fossil record does exist. The approach works well if you don't try to extract too much detail from it. So in what follows I will restrict myself to the major messages that come out of this work.

The solar system formed about 4.5 billion years ago. There is evidence for the activities of living organisms (as microbial fossils) in terrestrial rocks that are 3.5 billion years old. Life may have evolved before that (it certainly arose very quickly on the embryonic Earth), but calculations from study of the craters on the Moon suggest that up to about 3.8 billion years ago, the Earth-Moon system was subjected to gigantic asteroid impacts. These would have been large enough to release sufficient energy to re-sterilize the Earth's surface if any life had evolved. These cataclysmic impacts stopped about 3.8 billion years ago, and the Earth's surface stabilized sufficiently for life to evolve and for the first bacterial-product fossils to be laid down 3.5 billion years ago. Bacteria themselves don't make good fossils, but photosynthetic bacteria precipitate calcium salts into big characteristic mounds that do make good fossils.

After this there was a period of 1.5 billion years during which early bacteria continued to evolve before the higher organisms (called *eukaryotes*) emerged from their bacterial ancestors about 2 billion years ago. This critical event was followed by 1 billion years of eukaryote evolution (as single-celled creatures) before plants, animals, and fungi began to diverge from each other. So, finally, the major kingdoms of higher organisms have been separate from each other for the past 1 billion years. There is still uncertainty about the exact sequence of emergence of the major kingdoms, probably because of the effect of variable rates of evolution between the different groups (and, indeed, between the different molecules analyzed). The best available calculations indicate that plants, animals, and fungi last shared a common ancestor about 1 billion years ago; that recognizable animals originated about 800 million years ago; and that animals and fungi are each others' closest relatives. The sequence that emerges is that plants arose from the common eukaryote ancestor 1 billion years ago, then a joint fungal–animal line continued for another 200 million years until the animals left home 800 million years ago.

It's been argued that the oldest fossils found to date (which are about 650 million years old) are actually lichens rather than worms or jellyfish. This, however, is a hotly disputed interpretation. Fungi must have been around at that distance in time because from rocks about 570 million years old, we begin to get evidence (mostly in the form of spores because these seem to make good fossils) for all the major groups of fungi that exist today. Very early in evolution, intimate associations between fungi and plants occurred. Almost all land plants of today form cooperative associations with fungi. Plant roots are infected with fungi, which contribute to the mineral nutrition of the plant and can benefit plants in a variety of other ways. This association is the mutually beneficial mycorrhiza (fungus-root) that is described in chapter 4. This cooperation would have eased, if not solved, some of the most difficult problems the first land plants faced as they emerged from the primeval oceans. Some of the oldest (about 400 million year old) plant fossils contain mycorrhiza structures almost identical to those that can be seen today. It's now generally thought that the initial exploitation of dry land by plants about 430 million years ago depended on the establishment of cooperative associations; between fungi and algae (lichens) on the one hand, and between fungi and emerging higher plants (mycorrhizas) on the other. Fungi were crucially important in the shaping the ancient ecosystem.

Amber is good at preserving soft-bodied organisms like fungi. Fungal spores have been found in amber that is about 220 million years old. Several of these spores are almost identical to fungi existing in the present day. This is pretty remarkable. When they were trapped in the resin that hardened into amber, all of Earth's land masses were combined into one supercontinent (called Pangaea), birds were only just beginning to evolve, and flowering plants would not appear for another 100 million years! Fossils like this (and others) show that the characteristic fungal structures seen today arose long, long ago and have been maintained for enormous periods of time. One of the experts put it this way: ". . . the history of fungi is not marked by change and extinctions but by conservatism and continuity."

Probably the most remarkable find reported so far is amber containing the remains of two mushrooms that can actually be identified because they are so similar to existing mushrooms. But the amber is 90 to 94 million years old. Before the age of mammals, when dinosaurs still ruled the Earth, there existed mushrooms almost the same as those existing today. Mushroom fungi first evolved about 200 million years ago, but the mushrooms we see around us when we trek through the forests now are pretty well identical to mushrooms in the undergrowth through which dinosaurs trekked 50 or 100 million years ago. They survived whatever cataclysm brought extinction to the terrible reptiles. They've seen the mammals evolve to a primate that calls himself *sapiens;* there's no reason to doubt that they'll still be around when all the primates are dead and gone.

If a final reckoning of life on Earth is ever written, the fungi will figure from first to last. Arguably the first higher organisms to evolve, they in a sense gave rise to plants and animals, maybe 2 billion years ago. Later, they enabled plants to invade the land to start terrestrial development of planet Earth, helping the plants to shape nature as we know it today. We would not be here without fungi because their interventions and contributions have been crucial in the development of life on land to the point where it could support larger animals. And all the while, the fungi themselves were so well adapted to even dramatically changing environments that their own evolution was slow and relaxed.

Today fungi range from among the smallest to the largest individuals on Earth. The yeasts are among the smallest, yet we use them to make enough alcohol every year to refloat the Titanic. In the Malheur National Forest in eastern Oregon, where there is little more than a curious mycologist and the breeze to disturb the leaves, a monster dwells. It is eating the forest. The monster covers an area of 890 hectares, weighs about 150 metric tons, and is at least 2,400 years old. This makes the monster among the largest, heaviest, and oldest living things known on this planet, but it's no alien creature. You must have guessed it by now; the monster is a fungus, a clone of a tree root pathogen known as *Armillaria ostoyae.*

9

Birds Do It, Bees Do It, Even Educated Fleas Do It: But Why?

Look around. There are living organisms everywhere. We have a special word for it: *biodiversity*. We even have international conventions about it. Article 2 of the 1988 Convention helpfully tells us that "Biological diversity means the variability among living organisms from all sources including, inter alia, terrestrial, marine and other aquatic systems and the ecological complexes of which they are a part; this includes diversity within species, between species and of ecosystems." All that boils down to the life-support system of planet Earth on which we depend for the air we breathe and the food we eat. We worry about it, of course, because we have recognized over the past few years that human activities are destroying ecosystems. We make deserts out of fertile plains, turn tropical forests into pastures, and pollute entire oceans. Then we worry about it.

One important, and useful, outcome of the worrying has been the recognition that a problem might exist. Another is the realization that we need to know what the situation is now before we can measure how much change occurs over the next year, or decade, or century. In other words, we've got to start with the question: "just how diverse is biology?" We might approach an answer to that question by counting how many names we have for different organisms. We have been naming things ever since we've been able to communicate, but really scientific naming has been going on only since Carl Linnaeus established the first widely agreed upon set of rules in the middle of the eighteenth century, just about 250 years ago. We give a species a particular name in order to catalog the accumulated knowledge about it and provide a means to recognize it when it is found again. Amazingly, there is no single catalog yet in existence of the species we already know about. Lists and catalogs cost money, and nobody has yet compiled anything like an index of species known in the world. It would be a pretty large undertaking. About 20 thousand new species are named each year, and the total we currently know about is about 1.5 million. Perhaps this sounds like a lot, but the best guess at the moment is that we have given a name to little more than 10 percent of the species that actually exist. The United Nations Environment Program has estimated that there are 15 million species of organisms on planet Earth right now. If that's true, then, at our current rate of progress, it will take another 700 years to find and name each one that's unnamed at the moment.

Knowing how many species there are in nature is important to appraisals about the impact of human activities on the natural environment. Even more important is the judgement that has to be made about whether we should change our activity to alter that impact; and how the change is to be accomplished (and at what cost, and to whom) if we decide to change. These considerations are crucial to our future on the planet, but they are outside the scope of this book. I have ventured down this line of thought for a different reason. A useful approximate definition of a species is that it is a collection of individuals who are able to interbreed. This implies that two different species cannot interbreed, whereas members of the same species can interbreed. In other words the whole definition of *species* is bound up with the process of sexual reproduction. There are exceptions, however. Some genuine species hybrids are possible. But these are exceptions, and the ability to reproduce sexually is a very reliable guide to species identity. In a very real sense, therefore, sexual reproduction is the process that drives biodiversity. There are 15 million organisms on Earth because of the numerous rounds of sexual reproduction that have taken place in the 3.8 billion years that life has been evolving on the Earth. Obviously, sex is important to living things, but why? Why do organisms reproduce sexually? Like every other attribute of living creatures, sexual reproduction must have evolved, so there must have been a time before it evolved when the simple creatures of the day didn't have sex. So when and why did sex begin?

The answer to when is reasonably straightforward, it must have been a long, long time ago; but the answer to why is much more difficult because it leads us into deep philosophical waters. The difficulty is that sex involves the genetic material of one organism combining with the genetic material of another. Inevitably, sex decreases the representation of an individual's genes in the next generation, so what is the selective advantage that enabled it to evolve and become such a dominant force? Let's leave the "why" to one side for a moment and try to find out "when".

Plants and animals make pretty good fossils, and a lot of people expect fossils, particularly fossils of animals, to answer most questions about evolution. And, yes, you can find evidence of sexual activity in fossilized things that sound as though they are very old. For example, several museums have pieces of amber that contain pairs of insects that were mating when they were trapped in the resin that became amber. These may be 40, 50, or 60 million years old: old, but not that old. The earliest fossil dragonflies, which were flying around about 300 million years ago, had complex male and female genitals. And we can see the sex business even further back than that. Four hundred to 500 million years ago there were small, shelled creatures called ostracods, related to the *Daphnia*, or water fleas we use to feed goldfish, which had clearly different males and females. The females had large brood pouches, and the males had a double penis up to a third of the animal's body length, which was very extravagant sexual development half a billion years ago. But it seems to have paid off. These creatures colonized every aquatic environment on the planet and have survived every cataclysmic extinction to occur in that half billion years, and they're still going strong today. That sort of persistence is evidently one of the advantages of sex: it helps a species to survive. At the other extreme, another advantage of sex is that it promotes variety of life. And we can see this because during the same 500 million years that ostracods have been bumbling around, other lines of evolution have produced fish, amphibians, reptiles, birds, mammals, and humans.

Obviously we have to go much further back in time than a mere 500 million years, so let's try another tack. In chapter 8 we saw that there is evidence for fossil microbes in rocks that are 3.5 billion years old. They don't look like much more than circular or cylindrical blobs of stone between one and five thousandths of a millimeter long, but they are thought to be the oldest bacteria yet found. Some of these blobs have constrictions along their length that have been interpreted as indicating that the cells were in the process of dividing, splitting into two to reproduce themselves, at the time they were killed. Bacterial cells are too simple to have enough organized structure for us to recognize differences that might be sexual, so to make a guess at when and why cells like this started to share their genetic material with a mate, we will have to look at what happens in present day bacteria.

Sharing genetic material requires the physical mixing of genes from two different sources. In present day animals and plants, it's eggs and sperms and pollen that bring together genes from the mother and father. Bacteria don't have such highly evolved machinery, but they do seem to have a more basic, down-to-earth relationship with their surroundings. So basic that they can absorb genes directly from the fluid they are living in. Viruses can also carry bacterial genes from one bacterium to another. Both of these events can result in a bacterium having a mixture of genes from two sources: the essence of sexual reproduction. But the process in bacteria that is closest to a sexual one is when two bacterial cells join together and genes from one flow into the other. This is called *conjugation*. The remarkable thing about this is that it's not controlled by bacterial genes, but by an alien genetic element called a plasmid. Plasmids are stripped-down parasites. They don't make cells of their own; instead, they live inside other cells, mostly bacteria. They are all quite simple, and the very simplest ones only have genes for processes needed for their own reproduction. But some also organize cell-to-cell contact, that is, conjugation, between their host cells. What's in it for the plasmid is that conjugation allows the plasmid to pass like an infection from one bacterium to another of the same generation. Plasmids that do not organize conjugation are trapped within their host and can only get into a new cell when their host cell reproduces. So conjugation can open up the whole population to infection, giving the plasmid an enormous biological advantage. What's in it for the bacterium is that occasionally the plasmid genes get stuck onto the bacterial genes, and the plasmid transfers these to the other bacterial cell. That second bacterial cell then ends up containing genes from two sources: its own genes and other bacterial genes from its partner. The genetic information is mixed, and the necessary prerequisites of sexual reproduction are established. These sort of events occur in all groups of bacteria we know of today, even those that are thought to represent the most ancient and primitive forms.

So it could be that sexual reproduction originated when bacteria took advantage of an infection mechanism evolved by plasmids originally for their own purposes. Remember that they had a long time to achieve this. Bacterial evolution occupied a period of about 1.5 billion years before the higher organisms arose from their bacterial ancestors about 2 billion years ago. And the first higher organisms then had another billion years to perfect and adapt the process even before plants, animals, and fungi started to separate from each other about 1 billion years ago.

Today, plants, animals, and fungi undergo sexual reproduction using essentially the same biochemical tools. So if they all have a common process, it's reasonable to assume that it *is* common because the process evolved before evolution separated these three higher Kingdoms. This must mean that sex was first practiced by those early organisms that were neither animal, plant, nor fungus but some primitive compromise between all three. Sex reared its ugly head that long ago. We've had one billion years of fooling around! Maybe it's about time we asked why. To deal with this I really need to talk about one group of organisms: fungi. This is not to say that the arguments don't apply to the rest, they certainly do; but it is easier to channel this sort of thinking if we have a particular lifestyle in mind.

Most fungi produce spores that result from a nonsexual process. It might be called asexual reproduction or vegetative reproduction. Whatever it's called, it's just a matter of fungal cells breaking up in some organized way. This makes little parcels of living material that might be protected in some way (like a wall that prevents them from drying out) and usually are adapted to being dispersed (like having long thin processes that help dispersal in water currents). We have probably all seen moldy food or mildewed clothing covered in a green, black, or brown powder. The powder is made up of the spores of the mold, and even small quantities of substrate can generate vast numbers of spores. These spores are extremely effective in dispersing the organism. They are in the air all around us, in the waters of streams, rivers, and reservoirs, and among the dust that accumulates on surfaces in houses, cars, and yards. If one of these spores lands on the surface of a potential source of food for the fungus, then the spore will germinate and a new fungus colony will grow. The genes of this daughter colony will be the same as the genes of the parent colony unless a mutation has occurred in one or more of the genes, either when the spore was first produced or while it was being dispersed. A mutation will change the function of a gene, usually making it work less well. But that's a value judgement. Literally a life and death value judgement, because if the mutation causes a really adverse change, then the organism will be crippled and might die. Something that is unsuited to the present environment will die off. This is evolutionary selection in operation. By selecting against genes that have a poor performance, evolution ensures that the genes that remain in the living population are among the best that have been tried so far in the present environment. That's why most mutations are bad; the good ones have occurred before and have been selected to operate in the present environment. I keep emphasizing the present environment because circumstances can change and the genes that are so-called good now, say in middling temperature and humidity, may not be equally good if the temperature or humidity increased. That's natural selection: make the genes work in the conditions where the organism lives. Mutation keeps generating variants. If the conditions don't change, the mutations are most likely to lose. But if the conditions change some of the mutations might be winners. The rate at which mutation occurs is usually rather low. A representative rate to keep in mind is around one in a million; there are higher rates, there are lower rates, but one in a million is a fair average. So if a fungus produces about one million spores, there's a fair chance that at least one carries a mutation.

Look again at that moldy food or mildewed cloth: there are tens, hundreds, thousands of millions of spores. And tomorrow there will be millions more, and so on. Such colossal numbers of asexual spores are routinely produced that you might expect that there are a sufficient amount of them for mutation alone to provide all the variation that evolution might need for its operation. We can see this point now with any moldy material, but remember that we are searching for an argument that might have operated more than 1 billion years ago, and this point would have been just as valid then. Turn the clock back to a time when the highest organism alive was a primitive single-celled plant-animal-fungus creature. It may have been primitive, but the population must have been numbered in many billions. Let's assume that it was so primitive that it took a week to duplicate itself. That sounds pretty weak for an organism that is already, presumably, competing pretty well with all the bacteria that already existed then, but we can work with it. Imagine we have one such organism on January 1, we'll have two on January 7, four on the 14th, and eight on January 21. By the end of June we'll have 33 million, and there'll be 2000 million million of these little wimps by the end of their first year. And, though they may not know it, they've got a billion years ahead of them to keep on doing the same old thing. This may not be a very likely scenario; these organisms are weak. So a few will probably die, but it makes the point that simple duplication can make enormous numbers of the simplest organisms quite quickly. And there is an awful lot of time for it to happen. So you can easily see that even if mutation is only a one-in-a-million thing, every possible mutation in every possible gene will occur in the population at some time. But we know that the ancestors of these organisms (the bacteria) were already investing some effort in sexual reproduction. And we also know that the descendants of these organisms (present day plants, animals, and fungi) invest enormous resources of time and effort in their much more complex sexual rituals (and many animals have abandoned asexual reproduction altogether). Clearly, sex must have distinct selective advantage because it has emerged alongside, and in some cases to supplant, asexual reproduction. Importantly, the selective advantage must be one that applies to the most primitive organisms and to those of the present day.

The crucial contrast between sexual and asexual reproduction is that sexual reproduction brings together genetic material from different individuals. In asexual reproduction, only the genes of one individual are multiplied as it replicates and divides. There must be something about bringing together genetic material from different individuals that provides immediate evolutionary advantage to the organism in which it happens. It doesn't have to be much of an advantage; any advantage will give it an edge in the battle for favorable selection. If the two individuals whose genetic material is brought together are different in some way, then their differences can be rearranged during the sexual process. So the sexual mechanism enables new combinations of characters to be created in the next generation for selection. This is the most usual explanation for sex, namely that it promotes gene variability through what is called *outcrossing*, and if that is put together with the expectation that variability is needed for the species to evolve to deal with competitors and environmental changes, then sex is seen as an all-around good thing to do. It's got to be admitted that plenty of evidence exists to show that out-crossing certainly does promote variability, and that asexual organisms change only very slowly over time. Both of these lines of evidence seem to support the view that variability in the population enables the organism to change in order to survive ecological and environmental challenges.

The problem with such a line of argument, however, is that it is a group selectionist interpretation. It argues that variation generated in individual organisms is advantageous because it benefits the group or population to which the individual belongs. This goes against the current fashion in evolutionary thought because current theory emphasizes instead that selection acts on individuals. It follows, therefore, that any feature that is argued to be advantageous in selection must be advantageous because it benefits either the individual itself or its immediate offspring. On those grounds, this idea does not adequately explain why sex is an advantage to the individual organism. Another idea is that since mutations are mostly damaging, simply bringing together genes from different individuals in the prelude to sexual reproduction will ensure that any mutation-damaged gene in one set will be masked by its nonmutant counterpart in the genes of the other parent. This would, potentially, be of immediate advantage to the individual that contains the two sets of genes, but the real advantage seems to be that damaged DNA caused by mutation or faulty duplication in one set of genes can be repaired by comparison with the normal one provided by the other parent. That seems to be the really crucial advantage of the sexual process. An opportunity to repair damage, providing immediate advantage to the individual and long-term advantage to its offspring. It's fun as well, although oddly enough, *fun* doesn't appear to have been rated very highly in terms of evolutionary significance.

Before any advantage can be gained, though, the genes from two individual organisms must be brought together. So the first step in any sexual cycle is a cell fusion, or at least a cell-to-cell contact. The process involves breakdown of two walls before the two separate cells can unite, but even before that stage, the cells must be able to recognize each other. Sex starts with those steps, and evolution of the mechanisms that control recognition and union between two separate cells was a cardinal operation in the whole procedure. For the individual cell (and remember, it was at that stage of evolution that the evolution started) the problem is how to regulate cell unions so that the genetic advantages can be realized without hazard. There are contradictory requirements. Maximizing the advantage of sexual reproduction requires that the genes are as different as possible. On the other hand, safety for the cell requires that if cell contents are to mingle, they must be as similar as possible. The danger at the cell level is that cell fusion carries the risk of exposure to contamination with alien genetic information from things like viruses or plasmids. They may not be damaging to the cell that contains them (because it's adapted to their presence), but when it fuses with another cell, the second parent may suffer badly.

Protection against alien DNA is provided by a particular family of genes that only allow fusion with cells that have similar or identical genes in their corresponding copy of the family. These are called vegetative compatibility genes. They determine whether two cells are sufficiently compatible to fuse. In complex organisms like animals and plants, gene families of this sort exist, but they have been increased greatly in number and have now branched out to include features that determine gender: they make males express male characters, and females express female characters. In simple organisms like fungi, the vegetative compatibility genes are the closest thing they've got to a self–nonself recognition system. When the colony edges of two fungi grow to meet each other, the leading hyphal cells may mingle without interacting, or hyphal fusions may occur between their branches. If the colonies involved are not compatible (that is, if they have different vegetative compatibility genes) the cells that first fused are killed. Vegetative compatibility prevents successful fusion except between fungi that are sufficiently closely related to belong to the same family group of vegetative compatibility genes. In fungi it is these genes that determine individuality. That's important in evolution because selection operates on individuals. When the vegetative compatibility genes do allow individuals to exchange genes, there are other genes (called mating or breeding systems) which then come in to play to regulate sexual exchange. Primarily, these genes prevent mating between genetically identical individuals. Between them, these families of genes achieve the required balance between similarity and difference; the ideal compromise. We know about this from our own experience because we have genes that make males into males and females into females; we have two sexes. Fungi seem to have gone overboard in this direction because their mating type system can generate hundreds, even thousands, of different sexes. It's not entirely clear what the strategy for this peculiarity is; maybe there's more going on here than we realize.

The outcome of sex is the rearrangement of the genes that were brought together into new combinations packaged into the offspring spores that can be distributed into the environment. Many fungi produce fruiting bodies in which spores are produced and that both protect and distribute the spores. These are the truffles, puffballs, toadstools, and mushrooms. These are composed of different tissues, so fungi can quite reasonably be considered to be multicellular organisms. Like animals and plants, they are able to produce tissues comprised of cells that have different functions. Fungal developmental biology is alive and well, but we need to know much more about it. Tissue patterns in mushrooms are established very early in development. Mushroom embryology suggests that processes known to occur during animal embryo development have their analogs during mushroom development. Unfortunately, although all of these processes are well researched in animals (and increasingly in plants, too), that other great kingdom, Fungi, is being largely ignored. So while the occurrence of these processes during mushroom development may be inferred from indirect observations, there are very few specific examples of research aimed directly at understanding how fungal multicellular structures develop. Fungi are simple, but they are not primitive. They have several complex, sophisticated, and highly evolved relationships. They produce some remarkable structures. It all goes to show what can be achieved with simple tools.

10

The Cavalry Is Coming

Fungi to the Rescue

Human activities are making a mess of our one and only planet. We pollute our air. We pollute our water. We pollute our food. For short-term gain, we cause long-term damage. The only face-saving aspect of it all is that for most of the time our technology has been reshaping the planet, we have been unaware of the damage we have been doing. Like children playing a game, we have wrought a civilization with no appreciation of the harm our activities have been doing to the fabric and holistic life of the Earth. If there is anything special about the human animal, it is that we have the ability to set our own constraints. In nature, competition, predators, disease, and available resources constrain populations. It's our technology and our ability to devise it that makes us different from other animals. Strip us of our technology and send us naked into nature and we are not very special. Without the tools that enable us to punch above our weight, nature once more imposes on us a balance determined by outside forces. But our technology allows us to step outside the biological constraints and impose our will upon the world. This technology I'm thinking of here is not at the nuclear power and rocket science end of the spectrum. Once this primate discovered fire, a single human or ape-like creature could level a 1,000-year old forest into scattered smoldering embers. By inventing flint axes and arrowheads, he can inflict death and destruction on the most fearsome predators. Give him the wheel and a lever, and he will move mountains. Domesticate the horse, and the primate can roam over tens of thousands of miles in a single lifetime. When the primate invented agriculture, the climax vegetation stopped being forest and became whatever he wanted to grow, be it cereal grasses, tobacco, or poppies.

We have been modifying nature systematically since we lived in caves and chased mammoths. The only possible defense we can offer for our adverse impact on our environment is that for most of the time we were ignorant of not only the extent of the damage we were doing, but that we were inflicting damage at all. We know about it now, and our damaging activities cannot be excused, let alone denied. But we face conflicting demands. We now depend on continuation of our current technological activities. We can't consider the prospect of even slowing down our present industrial–commercial civilization because there are 6 billion of us. Admittedly, a lot of those people don't get their fair share of the resources available. But we would not be able to support a fraction of 1 percent of that population without the much reviled industrial–commercial machine we have created. We cannot live without it; equally, the planet cannot live with it, at least not in its present form. The paradox is that we cannot afford to stop and we cannot afford to continue. The resolution of the paradox is that we could change how our industrial–commercial machine works its magic for us. A compromise must be established, and over recent years, the name of that compromise has become sustainable development. Arising out of world environmental commissions and finally the U.N. Earth Summit in Rio de Janeiro in 1992, sustainable development has become a priority for the world's politicians and policy makers. Its foundation is the recognition that it is not possible to return to a natural lifestyle but that the unnatural lifestyle we have created for ourselves must be managed in such a way that future generations can continue to enjoy it. A commercial analogy summarizes it best: "the Earth must be managed in the way good managers would run a company if they were intending to pass a going concern on to their children."

Having recognized that a problem exists, the policy makers have tried first to estimate how much it will cost to deal with it. In Europe it's been estimated that improving waste management will cost 300 billion U.S. dollars, half that again to ensure good quality drinking water, and almost 400 billion U.S. dollars to counteract the greenhouse effect. The equivalent figures for these three programs in the United States are 270 billion, 100 billion, and 370 billion U.S. dollars, respectively. Estimates of other, this time global, costs are: over 100 billion U.S. dollars to protect the topsoil on agricultural land, over 30 billion U.S. dollars for renewing forests, 155 billion U.S. dollars to slow the rate of human population growth, 200 billion U.S. dollars to raise the efficiency of energy usage, and 100 billion U.S. dollars to develop renewable energy resources. These are the estimates of likely costs at the start of the program, but they are bound to rise.

As you can see from the nature of the programs listed, what we are facing is a biological problem. Humans, through a variety of their activities, are causing biological damage to their environment. If the problem is biological, then its solution can be biological too. We can see this in some of the programs that have already been started. We are used to hearing about renewable energy sources, for example. Water and wind-driven power generation seem to be at the top of this list, but biotechnological alternatives to fossil fuels could become important too. Now these might include materials like alcohol; produced, for example, by fermentation of some industrial waste product such as molasses. Sadly, after 25 years of development, Brazil and North America are still the only regions producing large quantities of fuel-alcohol, from sugar cane and maize, respectively. Unfortunately, even in these regions it is only tax credits that make fuel-alcohol commercially viable because oil prices are so low. However, tax considerations may encourage the use of fuel alcohols in aviation. The airline industry is one of the fastest growing sources of greenhouse gases at the moment because of its rapid expansion and dependence on kerosene produced from crude oil. A report in the London *Observer* for February 27, 2000, described how British Airways, Rolls Royce Aerospace, and several U.K. Government ministries were cooperating in a venture to assess whether planes can be fueled with a kerosene alternative produced by fermentation of straw and wood pulp. Although expensive alternatives at the moment, such fuels are likely to become viable alternatives as increasing taxes are imposed on oil-derived fuels following international agreements based on the principle that the polluter pays.

It may seem perverse to go to such extremes to replace one burning fuel with another, but it's not. Even direct burning of wood, or some more rapidly produced plant biomass that can be burned as a fuel, is better for our environment than burning coal or oil. Provided that appropriate steps are taken to minimize direct pollution from smoke and particles in the flue gases, it is more ecofriendly to burn wood than it is to burn fossil fuels because burning fossil fuels increases the atmosphere's content of carbon dioxide. Burning wood also produces carbon dioxide, but when something is burned that has grown over the past few years, the carbon dioxide that has been in the atmosphere during the past few years is simply recycled. So the removal of atmospheric carbon dioxide as the wood was formed and its sudden release as the wood is burned are just minor perturbations in what amounts to the present day carbon dioxide level of the atmosphere. When you burn oil or coal you are releasing into today's atmosphere carbon dioxide that was used in photosynthesis 300 million years ago. In effect, burning oil and coal short-circuits the natural carbon cycle, reinjecting into the atmosphere chemicals that were removed long, long ago. This disturbs the natural balance of the atmosphere and has resulted, as we all know, in the problem of the greenhouse effect.

Fungi could certainly help us cope with at least some of these problems. Since the problems we have created for ourselves are biological ones, there are many areas in which application of biotechnology might provide solutions, for example, in terms of improving food production, creation of renewable resources, and dealing with waste materials. Fungi can contribute a great deal to these solutions. There is no way that I can set out a comprehensive recipe for solutions here, but I do want to describe a few topics in which I believe current thinking patterns among the experts are too narrow and risk losing opportunities that the fungi offer.

The underlying food production problems are widely seen as resulting from the rapidly growing human population and the damaging environmental effects of today's most efficient agricultural production systems. The most uncomfortable statistic from which conventional experts start is the fact that approximately 40 percent of crop production worldwide is lost to pests and to pathogens, and to competition with weeds. The experts see a number of potential remedies in which biotechnology can be involved. Among these are the need to focus attention on the production of more food on the same area of land, with the aim of reducing the pressure that continues to expand agriculture into the wilderness or forest areas that are the natural reservoirs of biodiversity. Obviously, there is also a perceived need to reduce losses of harvested crops and so effectively increase crop yield; to replace conventional fertilizers and pest control agents (which are resource- and energy-demanding) with more environmentally friendly alternatives; and finally, to encourage replacement of environmentally damaging agricultural practices with more enlightened crop management techniques.

It seems to me that there is another statistic that deserves some attention. Very little of what agriculture produces is actually used. Our agricultural efforts produce more waste than anything else. This cannot be allowed to continue, yet it is a feature that rarely appears in the policy maker's deliberations. That's worrisome because the wrong decisions have been made before. The so-called green revolution in the 1960s involved the introduction of new high-yielding varieties of cereal crops like corn and wheat in order to increase the world food supply. Unfortunately, the new varieties required good irrigation and application of large amounts of chemical fertilizers and pesticides that were costly to produce and expensive to buy. Inevitably, only the richer farmers were able to benefit from this alleged revolution, and the increased applications of chemicals damaged the environment. The approach was completely wrong; there was so much misguided political conviction behind it that it took 30 years before the green revolution started to die away. We cannot continue to expect the Earth to produce more and more; what we must do is make better use of what the Earth already produces. Over 70 percent of agriculture and forest production is not useful or is wasted in processing. Just think of a field of flowing corn; only the seeds are used. All of the plants that you can see rustling in the breeze are waste and contribute to the 3 billion metric tons of cereal straw wastes we produce each year. The same astonishing level of wastage applies to every cereal crop. Or consider fiber production from sisal; the extracted fiber is only 2 percent of the sisal plant, and the other 98 percent is thrown away as waste: tens of millions of tons of waste every year. Palm and coconut oils represent only 5 percent of the total biomass grown on palm and coconut plantations; all the rest is waste. Sugar cane is better. As much as 17 percent of sugar cane plants emerges as sugar; only 83 percent of what is grown is wasted.

Apart from agricultural and forest wastes, there are growths of weeds, especially in Africa, that cause problems. Water hyacinth is the prime example. It is widespread on the surface of most tropical African rivers, dams, and lakes. In some places it produces so much biomass that it prevents navigation of water channels and damages fisheries as well as transport. Growth of water hyacinth consumes production resources to an astonishing degree: several African countries produce millions of tons of water hyacinth biomass. All of these waste materials squander the materials and efforts put into growing the crops to a scandalous degree, and they become an embarrassment in their own right. They are wastes for disposal by burning on site, burying, or dumping in landfills. So, not only are we making barely marginal use of the primary production of the Earth, but we are also creating a waste disposal problem for ourselves in the same process.

There's got to be an alternative approach, and maybe we should start by recognizing that these agricultural waste materials, which are abundant and readily available in every corner of the world, especially in tropical and subtropical regions, are resources because they are potential substrates for mushroom cultivation. Concerted use of agricultural and similar wastes could produce millions of tons of edible mushrooms for table use; and millions of tons of organic fertilizers from the spent composts. The approach can be applied at the cottage industry or peasant farmer level or at the more industrial level. The mushroom industry already has most of the technology that is needed for this. What is required is proper promotion and active support by the private sector and governmental organizations. Bear with me for a moment, and I'll explain why mushroom fungi are particularly useful and suggest a range of different things we could use the mushrooms for.

The great significance and promise of mushroom fungi lies in the fact that they belong to a group of fungi that can degrade woody materials. They have spent 100 million years or so evolving the ability to secrete enzyme complexes that are able to convert various woody materials into nutrients suitable for fungal growth. Not just timber, but stems, leaves, roots, and all the other large and small bits of vegetable matter that accumulate as plants grow and die. They are almost unique in their ability to digest woody material completely. A very few bacteria can make some effort in this direction, but these mushroom-related fungi are responsible for degrading by far the greatest proportion of waste plant materials in nature, so why not utilize that natural specialist activity to deal with our agricultural and other wastes? It seems perverse, to me, that various laboratories around the world are spending millions of dollars trying to genetically engineer bacteria for a job the fungi have been doing since before the dinosaurs appeared. Microbial degradation of woody materials is difficult by design, that is, by the design of the plants! The chemicals that make up wood contain several phenol-derived compounds with the specific purpose of protecting the plant tissues against microbial attack. Remember that wood is made up from the walls of plant cells, and the phenolics are put into those walls to protect the plant cells from invasion by microbes. They are natural antiseptics. The fact that so few microbes can digest wood is a measure of how effective the plants have been with all that evolutionary self-protective effort. The fact that the fungi have breached those plant defenses is a tribute to the fierce competitive edge of the fungi. They have found a way to use a resource that is resistant to attack by other organisms. They do this by producing enzymes that produce and use highly active oxygen molecules. To all intents and purposes the fungi rip the wood molecules open in a precisely controlled burning reaction. These are the allies we should use!

All right, so on biological and chemical grounds the more advanced fungi, especially the mushroom fungi, are the ideal candidates to degrade the waste vegetation that we produce through our agricultural activities and which, like water hyacinths, grow as weeds in regions that are not cultivated. But their usefulness is not limited to getting rid of waste vegetation. Not only do we have large amounts of wastes, but some of them are polluted with pesticides that are chemically similar to the complex phenolic compounds found in wood. Because the fungi can decompose the wood, they can also be used to degrade environmental pollutants in soils and in liquid effluents, such as industrial wastewater discharges such as those produced by the paper pulp industry, but especially pesticide-contaminated wastes. Many of the mushroom fungi and their relatives can degrade environmental pollutants such as chlorinated biphenyls, aromatic hydrocarbons, dieldrin, and even the fungicide benomyl. They don't just degrade these materials, leaving other possibly dangerous substances behind, but they completely mineralize the pollutant so that its chemical constituents are returned to the atmosphere and soil as carbon dioxide, ammonia, chlorides, and water. The oyster mushroom is particularly good at this sort of thing. We've recently found that several mushrooms degrade chlorophenols that have been commonly used as disinfectants and preservatives for several years, and pentachlorophenol (more commonly known as PCP) has been the most heavily used pesticide throughout the world. It has been used in the United States mainly as a wood preservative and enormous amounts (about 6 million kilograms each year) have been sprayed over vast areas of central China to kill the snails that carry the schistosomiasis parasite. Although it's illegal to use PCP in most countries today, the chemical is very persistent, and most of what was released in the environment is still there. It's toxic, cancer-inducing, and has been declared a priority pollutant for remediation treatment. The conventional remediation strategy for PCP-contaminated is excavation and incineration or landfilling. Such methods are expensive, obviously destructive to the environment, and ineffective for anything other than so-called point source pollution. Bioremediation is a very promising alternative, using biological systems for the environmental clean-up. And by far the most promising technique seems to be to use the spent mushroom substrates remaining after harvesting mushroom crops. Ironically, these are often discarded as wastes themselves, but the spent compost left after oyster mushroom cultivation does two crucial things. Firstly, it absorbs, immobilizes, and concentrates PCP so it can be transported away from the contaminated site. But secondly, it also digests PCP completely, providing an integrated approach to remediation. You also get a crop of mushrooms!

Mushroom cultivation is a common practice all over the world, and the idea that hazardous waste materials could have their pollutants removed and produce a mushroom crop at the same time is exceedingly attractive. But the idea is not free of problems. We have found that oyster mushrooms can concentrate the metal cadmium (a common industrial contaminant) to such an extent that by eating less than an ounce (dry weight) of the most contaminated samples, you would exceed the weekly limit tolerated by humans. Cadmium is so toxic that this situation could pose a public health hazard. There are no worries about conventionally cultivated oyster mushrooms. The point is that if the mushroom is grown on composts that might be mixed with industrial wastes (in remediation programs, for example), then it would be advisable to monitor the heavy metal contents before mushrooms are marketed for food.

Actually, the ability of mushrooms to absorb metals is relevant to yet another potential area of fungal biotechnology. More and more attention is being given to using fungal tissue to take up metal ions from solution. The aim of this is to remove polluting heavy metals (even radioactive ones) from effluents by using the adsorptive capacity of living or dead fungal tissue. The fungal wall makes this possible. It has evolved over the years to attract and accumulate metal atoms that might be needed for nutrition. It's a chemically reversible binding reaction, so the fungus can take up the metal from its own wall, probably in exchange for a hydrogen

From the entire field of canola shown in the photograph (top), only the oil from the seeds will be harvested. Like other agricultural crops, only a fraction of what is grown is actually used. But we could grow a lot of mushrooms on agricultural residues and convert wastes into food, animal feed, pharmaceuticals, and other products. Likely candidates are shown (bottom): clockwise, these are *Agaricus, Pleurotus, Lentinula,* and *Volvariella.*

atom. But the different metals have much in common, so a wall evolved to accumulate a common metal that is important nutritionally can also accumulate the less-common metals that escape into the environment from our industrial processes. These polluting metals can be removed from wastewaters or other liquids by passing the effluent through a column containing the fungal material. Not only can you clean up polluted effluents this way, but you can also recover precious metals from industrial wastes (like the silver from photographic waste solutions and the gold from electronic chip industries). Current practice is to use ion-exchange resins for metal recovery, but the effectiveness of resins is reduced by the (very common) metals calcium and magnesium, and they are not good at adsorbing minute quantities of metals from large volumes of effluent. These are exactly the situations that the fungal wall has evolved to cope with, and metal binding to fungal tissue is not greatly influenced by calcium and magnesium and is capable of accumulating metals from solution that are present in only trace amounts. Once metal ions have been adsorbed, it is commercially essential to release them so that the fungal material can be reused and the metal can be recovered. This can usually be done quite easily by treating the fungus with solutions of sodium bicarbonate or ammonium carbonate. This adsorption desorption cycle can be repeated pretty often. As you might expect, efficiency of metal removal differs between different fungi; it's presumably linked to the lifestyle and ecology of the fungus. I previously mentioned cadmium accumulation by oyster mushrooms; our experiments showed that 30 kilograms of cadmium per metric ton of mushroom could be recovered. This is in the same ballpark as commonly used exchange resins. Experiments with other fungi have given recoveries of just under 200 kilograms of metal per metric ton and that level of yield is at least twice the uptake capacity of the best exchange resins. The message seems to be that there's bound to be a fungus able to recover the specific metal from whatever metal-rich effluent that is disposed of. It's worth the research effort to find it.

If you are pouring some effluent through a bed or column of fungal material in order to remove metals, you'll find that the solution that percolates through will be clarified. This is because fungi are also very good at adsorbing insoluble particles from solutions. Most fungi can adsorb almost any particle with which they are challenged. Again, it's related to their lifestyle and ecology. In nature, fungi explore their surroundings in a search for food resources. When they find a resource, it sticks to their wall and they produce the necessary enzymes to digest it. That's fungal nature. If you exploit their ability to trap particles you find that they quite readily adsorb metal particles (like colloidal gold), elemental sulfur, insoluble sulfides, charcoal, clays, and even magnetite (so you can make magnetic mushrooms, if you like). Imagine using this ability to remove particles from waste streams to clarify them before discharge into rivers. Unfortunately, despite all the research devoted to the possible use of fungi to recover metal ions and remove particulates, no large-scale industrial process has yet appeared, even though the fungal process can perform better than most conventional chemical and physical techniques of effluent treatment. One reason why biological processes are consistently ignored is that chemical engineers are uncomfortable with anything that's really new. The industry is accustomed to handling established chemical–physical systems, the customers are confident about buying them, and the whole unimaginative cycle is closed off to biological innovation. The vision is there for those with eyes to see.

A few pages back, I suggested we must make better use of the Earth's primary productivity. We cannot continue to accept an agriculture that loses 40 percent of its production to pest and disease and then throws away more than 70 percent of what's left because the crop always represents so little of what is grown. Work through that little equation; we only take 60 percent of what we try to grow through to successful production, and we use, at most, 30 percent of that, so the bottom line is that we make use of no more than 18 percent of the Earth's primary productivity. Of the 6 billion people on the planet, 1.5 billion don't have enough to eat. Yet we congratulate ourselves on a worldwide agricultural industry that can manage a magnificent 18 percent efficiency. It's time we tried something else.

A good friend of mine, Shu-Ting Chang, who recently retired from the Chinese University of Hong Kong, has suggested that we should try a nongreen revolution that makes use of that part of the primary production that is not used for mushroom cultivation for mushroom production and importantly mushroom products.

Remember that in nature there are mushrooms everywhere. They grow on trees, they grow on grass. They grow in arctic snow, desert sand, and tropical rain forest. They have spent several hundred million years evolving to exploit every terrestrial habitat. More than 2,000 species of mushrooms are considered good to eat, but up to now, attempts have been made to cultivate only about 100 species. Indeed, despite the fact that mushroom cultivation is a worldwide industry (about equal in value to the coffee crop), only about 30 species are commercially cultivated, and only 7 are cultivated on what could be described as industrial scale. So there's plenty of scope: scope for finding mushrooms able to use any of the waste materials available. There is scope to operate on small-, medium-, or large-scale farming; or subsistence or high-technology farming. There is scope to cultivate mushrooms for new markets or old markets. Mushrooms are highly nutritious foods, but they are underexploited. Mushroom tissue could be used as a meat substitute in processed ready-to-eat meals; if it can be done with Quorn, it can be done with mushrooms. But it's not just a matter of mushrooms for the table. Mushrooms for industry makes sense, too. Mushrooms to be used in bioreactors for metal-recovery and effluent filtering. Mushrooms to be used as a source of biologically active compounds for use in the prevention and treatment of human disease. And then there's the used mushroom compost, which makes a good animal feed. Plant proteins are deficient in amino acids essential to human (and animal) nutrition. Fungi are not deficient in those amino acids, so a compost full of fungal cells is a better feed for the animals. Animals can't digest the woody parts of the plants they eat, but if a mushroom fungus is grown on woody, twiggy material, it digests the wooden parts, and when the animals eat it, they can make better use of a greater proportion of the feed. So on two counts, used mushroom compost can make good animal feed. It also makes good soil conditioner. If you can think of nothing better to do with it, plough it into landfill sites. It's got the structure and constitution that can only improve the soil, and it can digest those persistent organic chemicals and pesticides we use in industry.

The nongreen revolution offers employment, economic growth, and protection and regeneration of the environment. And the mushrooms taste good! It's all there waiting for us. Come on world. Wake up to mushrooms.

SOURCES

If you would like some further reading about fungi you might like to try the following books. This is not an exhaustive list, just some of my favorites.

The Advance of the Fungi, by E. C. Large (Jonathan Cape: London, 1940).

Mushrooms and Toadstools: A Study of the Activities of Fungi, by J. Ramsbottom (William Collins, 1953; Bloomsbury Books: London, 1989).

In the Company of Mushrooms: A Biologist's Tale, by Elio Schaechter (Harvard University Press: Boston, 1997).

Magical Mushrooms, Mischievous Molds, by G. W. Hudler (Princeton University Press: New Jersey, 1998).

Morel Tales: The Culture of Mushrooming, by G. A. Fine (Harvard University Press: Boston, 1998).

For more about biotechnology:

An Introduction to Fungal Biotechnology, by Milton Wainwright (John Wiley & Sons: Chichester, U.K., 1992).

For a very detailed account of how fungal development works:

Fungal Morphogenesis, by David Moore (published by Cambridge University Press: New York, 1998).

For recipes and guidelines about collecting:

Mushroom Magic. 100 Fabulous Fungi Feasts and Marvellous Mushroom Meals, by S. Wheeler (Sebastian Kelly and Anness Publishing Limited: Oxford, U.K., 1997).

And for general background information about the living world:

Biology, 5th Edition, by N. A. Campbell, J. B. Reece, and L. G. Mitchell (published by Addison Wesley Longman Inc.: Menlo Park, California, 1999).

SOFTWARE CREDITS

The word games at the end of the book were prepared using three excellent pieces of software:

Anagrams were created with *Anagram Genius* software (available for Windows or MacOS computers) from Genius 2000 Software, P.O. Box 395, Cambridge CB3 9PJ, U.K. Credit card orders: Tel: +44 (0)151 356 8000; Fax: +44 (0)151 357 2813; Email: agenius@genius2000.com; Web:http://www.genius2000.com/

The word search puzzles were made with a piece of freeware called *WordSearch* version 2.0, written for Windows 95, Windows 98, and Windows NT by Yuntong Kuo (ykuo@ma.ultranet.com).

For the crossword, I used *Crossword Compiler* Version 5.0 for 32-bit Windows together with *WordWeb Pro* thesaurus/dictionary, Version 1.6 for Windows 95 and Windows NT, both from Antony Lewis [E-mail: compiler@x-word.com; Internet:www.x-word.com/users].

ILLUSTRATION CREDITS

Chapter 1. Photographs kindly provided by Mrs. Jo Weightman.

Chapter 2. Photographs of blighted potatoes provided by Dr. H. A. S. Epton, School of Biological Sciences, The University of Manchester. Images of Dutch Elm Disease taken, with permission, from the website of the Coalition to Save the Elms, 2799 Roblin Boulevard, Winnipeg, Manitoba, Canada R3R 0B8. This is a nonprofit charitable organization dedicated to the stewardship of the elm and other forests in the urban environment. The website can be found at the URL http://www.savetheelms.mb.ca/.

Chapter 3. Taken from Fig. 67 in *Researches on Fungi*, by A. H. R. Buller. Longmans, Green and Co.: New York; 1931. Photographs of a dry-rot-infested basement provided by Dr. Ingo Nuss, Brunnenstraße 6, 93098 Mintraching-Sengkofen, Germany.

Chapter 5. Photographs of *Ganoderma* fruit bodies provided by Prof. S. W. Chiu, Department of Biological Sciences, The Chinese University of Hong Kong, Shatin, Hong Kong.

Chapter 8. Electronmicrograph of hyphae (of the mushroom *Pleurotus*) provided by Dra. Carmen Sánchez, Laboratory of Biotechnology, Research Centre for Biological Sciences, Universidad Autónoma de Tlaxcala, México.

Chapter 10. Photographs of *Lentinula, Pleurotus,* and *Volvariella* fruit bodies provided by Prof. S. W. Chiu, Department of Biological Sciences, The Chinese University of Hong Kong, Shatin, Hong Kong.

All other images taken and prepared by the author.

WORD GAMES

Now, you can try the word games on the next few pages, just for fun. You'll find everything you need to answer the puzzles in the text of the chapters.

H Q K P A R I S M U S H R O O M R R I
S A U R I C U L A R I A C E F K H M H
J E L L Y F U N G U S H G L N E L A E
A T L E V G T E C M A N A I D O S T R
C R S K N B R Y S M O M A G N U K O I
S S W C M T P H P L M N E M C A G I C
T U B E R N I I G U E H K I E I B L I
K P K G O I G N L Z O G R E Q K P O U
E U E T T N Q I U G O A O F Y H O H M
U M T A O S N F E L G S L L N H B P U
Z U K N E A B F S A A L P Z B P E G L
B E E F F G A N O D E R M A L L S A R
N M A T S U T A K E Y C M M A E H T D
M K O W A R T S Y D D A P M C U I R T
R E N Y T R E M E L L A S X K R A L R
Z S H I S T R U F F L E N F O O N F N
P O M C E T S X M S M M I A A T G F D
P T E A L L E I R A V L O V K U G A U
W O O D E A R R P P M B A O S S U D R

CULTIVATED MUSHROOMS

AGARICUS
AURICULARIA
BLACK OAK
BUTTON
CHAMPIGNON
ENOKI
FLAMMULINA
GANODERMA
HEDGEHOG
HERICIUM
JELLY FUNGUS
LENTINULA
MATSUTAKE
MONKEY HEAD
NAMEKO
OYSTER
PADDY STRAW
PARIS MUSHROOM
PHOLIOTA
PLEUROTUS
SHIANG-GU
SHIITAKE
TREMELLA
TRUFFLE
TUBER
VOLVARIELLA
WOOD-EAR

M Y C O S E S L S L I C H E N S S S R
E S R P Y Q A U W M Y C O R R H I Z A
N D C T U S C M U I Z I H R A T E M L
S C I O H I T O R E N N Z M Y L R M U
T T R C R G N E L E H E K R T O I S N
U N G A I S I O R M D E T N W L O P I
M L G G B G Z L S A M O U G D Y T T T
S A A A A A N B B U B B N E S S C U N
C C I M I S D U E P C I W A U S B N E
F L O M A R P E F A R S U R G E I E L
L P A M U N A E A B U C A I R X C T S
A O Y V P S I L R T E V D N O I I Y U
M R L E I O H T L G H T E T O B R H T
M I D E A C S R A I I C A R T M A P E
U A E E F S E T O N M L A L I F G O L
L H E P M E T P M O F R L P K A A D O
I S E R P U L A S A M K A U C K Y N B
N T C H A N T E R E L L E B S T L E H
A E R G O T O F R Y E D T R U F F L E

SSS WORDSEARCH

AFLATOXIN	DEATH CAP	MYCOSES
AGARICUS	ENDOPHYTE	OYSTER
AMANITA	ERGOT OF RYE	PORIA
ARMILLARIA	FLAMMULINA	QUORN
ASPERGILLUS	FLY AGARIC	RINGWORM
AZOLE	FUNGICIDE	RUST
BEAUVERIA	GANODERMA	SERPULA
BLIGHT	LENTINULA	SMUTS
BOLETUS	LICHENS	SOY SAUCE
BOTRYTIS	METARHIZIUM	TEMPEH
BUNT	MILDEWS	TRUFFLE
CHANTERELLE	MONASCUS	TUBER
CLAVICEPS	MUSHROOM	WINE
COMPOST	MYCORRHIZA	YEAST

For a real challenge, try these anagrams: all of the words and phrases used to make the anagrams appear exactly as they occur in the text of the book. The relevant chapter is indicated in the final column.

No.	Anagram	Lengths of Words in the Solution	Chapter
1	A flag I cry	3, 6	1
2	Cool summer clothing	8, 10	1
3	Fine, laughing council	14, 5	1
4	Coin geometry	5, 7	2
5	If my omission conquer	10, 2, 7	2
6	Code of superior brain	13, 6	3
7	I am so cool! Thirty-eight bicycles!	3, 7, 11, 7	3
8	Pigs in foul age	8, 5	3
9	Evolve fiercest orangutan up	6, 2-9, 8	4
10	Crazy filming hour	11, 5	4
11	Tasteful trance	4-6, 4	4
12	Incorporates a flirt	8, 4, 6	4
13	A remedial distinction	11, 9	5
14	The stuff of rioters	6, 2, 3, 6	7
15	Shady storms do warm up	5, 5, 9	7
16	Smoother musicians followed	1, 9, 2, 4, 9	7
17	Evolving to encounter	10, 9	8
18	Fantastic, ugly mess	6, 11	8
19	Enormous, important and intervening	6, 7, 11, 7	9
20	Workshop to a museum	4, 2, 2, 9	10

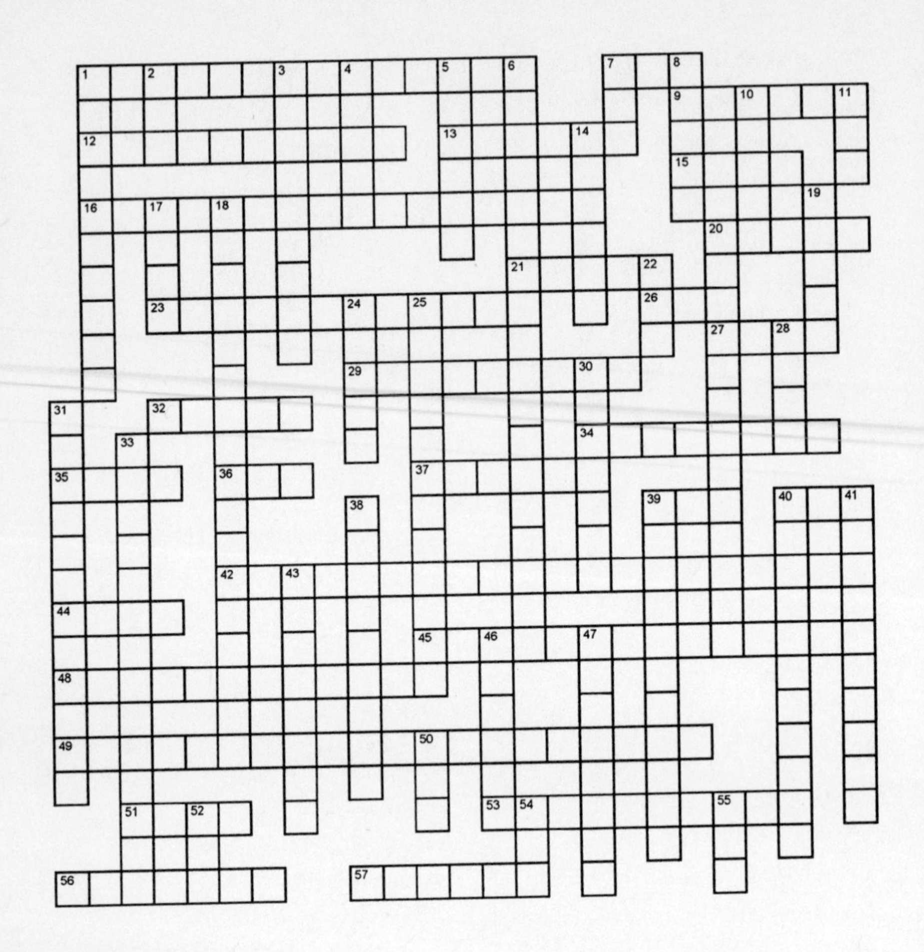

Across

1 Liking for damp (8-6)
7 Objective (3)
9 Arsenal (6)
12 Like a spasm (10)
13 Jewish republic (6)
15 Hard-shelled seeds (4)
16 Train wrecker (8,8)
20 Group of warships (5)
21 Third rock from the sun (5)
23 Poe's palace (5,2,5)

Down

1 Fungus freak (10)
2 Electrically charged atom (3)
3 Supplies (9)
4 Even (5)
5 Lichen reproductive structures (6)
6 Reported the advance of blight (9,9)
7 Pointed tool (3)
8 Miraculous food (5)
10 Shiny solid (5)
11 Valued for archery (3)

(continues)

Across *(continued)*

26 Lubricate (3)
27 Two and a half centimeters (4)
29 Most northerly point in Europe (5,4)
32 Led the Israelites from Egypt (5)
34 Water falling in drops (8)
35 Where U.S. dutch elm disease first found (4)
36 Metal vessel (3)
37 Hot art (anagram) (6)
39 Metal container (3)
40 Mist (3)
42 Cause of that lingering sparkle (5,15)
44 Sharp boundary (4)
45 The meaning (14)
48 Clear the ground (5,3,4)
49 A remedial distinction (anagram) (11,9)
51 Notion (4)
53 Wet trees (4,6)
56 Supervised (7)
57 Rice fermented by *Monascus* (3-3)

Down *(continued)*

14 Make certain (6)
17 First sailor? (4)
18 It inhibits response to antigen (17)
19 Departure from life (5)
20 He examined rusty wheat (6,7)
22 Spicy (3)
24 Fifth kingdom (5)
25 Actinomycete antibiotic (12)
28 Basic structural and functional unit (4)
30 An associate (7)
31 Variety of life (12)
33 I bargained to do (anagram) (14)
38 Elementary organization (9)
39 Animal sterol (11)
40 Adaptable quality (11)
41 Nursery (10)
43 From broad-leaved trees (4,4)
46 Kindling (6)
47 Sufferers (8)
50 Month (3)
52 Ovum (3)
54 Built to special order (3)
55 Organ of sight (3)

SOLUTIONS TO WORD GAMES

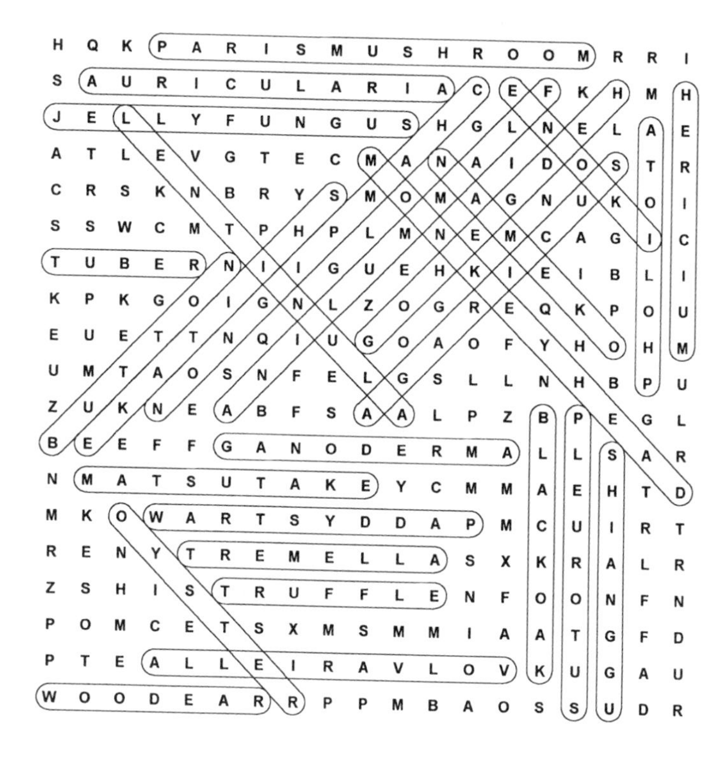

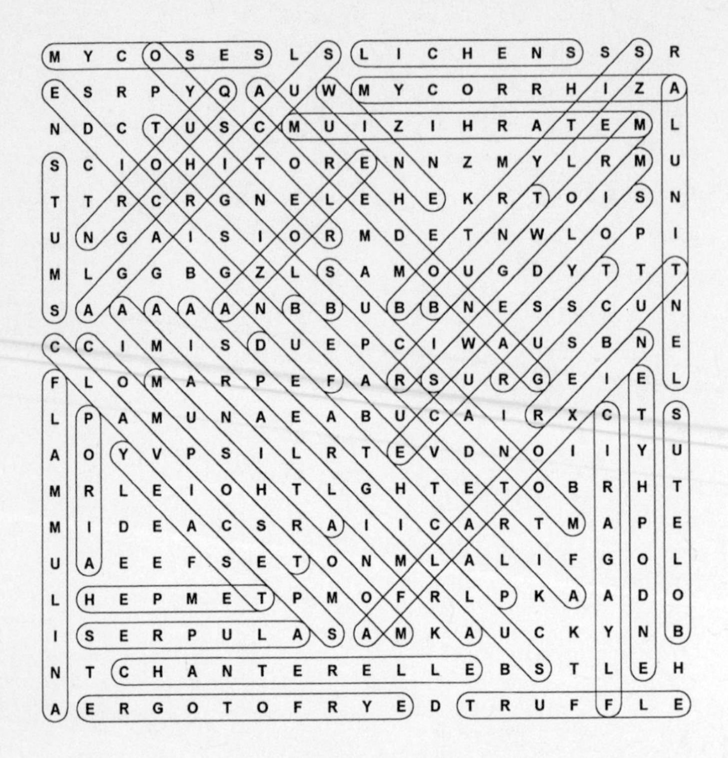

ANAGRAM SOLUTIONS

No.	Anagram	Solution
1	A flag I cry	fly agaric
2	Cool summer clothing	mushroom collecting
3	Fine, laughing council	hallucinogenic fungi
4	Coin geometry	Tiger economy
5	If my omission conquer	Commission of Enquiry
6	Code of superior brain	Carboniferous Period
7	I am so cool! Thirty-eight bicycles	The British Mycological Society
8	Pigs in foul age	spoilage fungi
9	Evolve fiercest orangutan up	fungal cooperative ventures
10	Crazy filming hour	mycorrhizal fungi
11	Tasteful trance	leaf-cutter ants
12	Incorporates a flirt	tropical rain forest
13	A remedial distinction	traditional medicines
14	The stuff of rioters	fruits of the forest
15	Shady storms do warm up	Paddy straw mushrooms
16	Smoother musicians followed	a selection of wild mushrooms
17	Evolving to encounter	convergent evolution
18	Fantastic, ugly mess	fungal systematics
19	Enormous, important and intervening	United Nations Environment Program
20	Workshop to a museum	Wake up to mushrooms

1 MOISTURELOVING
1 MYCOLOGIST
2 ION
3 RESOURCES
4 LEVEL
5 ISIDIA
6 GARDENERSCHRONICLE
7 AIM
7 AWL
8 MANNA
9 ARMORY
10 METAL
11 YEW
12 CONVULSIVE
13 ISRAEL
14 ENSURE
15 NUTS
16 LENTINUSLEPIDEUS
17 NOAH
18 IMMUNOSUPPRESSANT
19 DEATH
20 FLEET
20 FLEECE
21 EARTH
22 HOT
23 HOUSEOFUSHER
24 FUNGI
25 STREPTOMYCIN
26 OIL
27 INCH
27 ICECONTAINS
28 CELL
29 NORTHCAPE
30 PARTNER
31 BIODIVERSITY
32 MOSES
33 BIODEGRADATION
34 RAINFALL
35 OHIO
36 PAN
37 THROAT
38 MOLECULAR
39 CAN
39 CHOLESTEROL
40 FOG
40 FLEXIBILITY
41 GREENHOUSE
42 ETHYLPYROCARBONATENE
43 HARDWOOD
44 EDGE
45 INTERPRETATION
46 TINDER
47 PATIENTS
48 SLASHANDBURN
49 TRADITIONALMEDICINES
50 MAY
51 IDEA
52 EGG
53 RAINFOREST
54 ARK
55 EYE
56 MANAGED
57 ANGKAK

INDEX